RENEWABLE AND ALTERNATIVE ENERGY RESOURCES

Selected Titles in ABC-CLIO's
**CONTEMPORARY
WORLD ISSUES**
Series

For a complete list of titles in this series, please visit
www.abc-clio.com.

Books in the Contemporary World Issues series address vital issues in today's society, such as genetic engineering, pollution, and biodiversity. Written by professional writers, scholars, and nonacademic experts, these books are authoritative, clearly written, up-to-date, and objective. They provide a good starting point for research by high school and college students, scholars, and general readers as well as by legislators, businesspeople, activists, and others.

Each book, carefully organized and easy to use, contains an overview of the subject, a detailed chronology, biographical sketches, facts and data and/or documents and other primary-source material, a directory of organizations and agencies, annotated lists of print and nonprint resources, and an index.

Readers of books in the Contemporary World Issues series will find the information they need to have a better understanding of the social, political, environmental, and economic issues facing the world today.

RENEWABLE AND ALTERNATIVE ENERGY RESOURCES

A Reference Handbook

Zachary A. Smith and Katrina D. Taylor

CONTEMPORARY WORLD ISSUES

A B C CLIO

Santa Barbara, California
Denver, Colorado
Oxford, England

Library of Congress Cataloging-in-Publication Data
Smith, Zachary A. (Zachary Alden), 1953–
   Renewable and alternative energy resources : a reference handbook / Zachary A. Smith and Katrina D. Taylor.
       p.   cm. — (ABC-CLIO's contemporary world issues series)
   Includes bibliographical references and index.
   ISBN 978-1-59884-089-6 (hard copy : alk. paper)
   ISBN 978-1-59884-090-2 (ebook)
   1. Renewable energy sources—Handbooks, manuals, etc.   I. Taylor, Katrina D.   II. Title.

TJ808.3.S65   2008
333.79′4—dc22

                                                        2008016625

12   11   10   09   08   1   2   3   4   5   6   7   8   9   10

ABC-CLIO, Inc.
130 Cremona Drive, P.O. Box 1911
Santa Barbara, California 93116-1911

This book is also available on the World Wide Web as an eBook.
Visit abc-clio.com for details.

This book is printed on acid-free paper ∞
Manufactured in the United States of America

*This book is dedicated
to the many family members
and friends who provide me
with continuing support,
encouragement,
and hope for the future.
You know who you are.
—K. D. T.*

*For posterity.
Who hasn't done a thing for me.
—Z. A. S.*

Contents

Preface

With a focus on renewable and alternative energy, this book addresses issues that are vital for the question on the minds of scholars, policy makers, and society at large: What energy source will meet the growing needs of an expanding population with the least negative consequences for the environment and the economy? The purpose of this book is to provide the foundational information necessary to address this question and for readers to:

- Identify the array of renewable and alternative energy technologies available and under development;
- Become familiar with the scholarly and political arguments for and against the worldwide adoption of these technologies; and
- Understand the current debate for how best to make widespread adoption possible, including an understanding of different policy mechanisms for encouraging renewable energy production and use.

The first chapter provides an overview of energy sources used throughout the world. It documents the unquestionable dominance of fossil fuels in the energy market and contains the most up-to-date estimates of current and future worldwide energy use. The main purpose of the chapter is to furnish readers with detailed descriptions of the various renewable and alternative energy technologies that are available and in development. Each type of renewable and alternative energy is explained under the subheadings of History and Uses, Process of Utilization, Advantages, and Prospects for Meeting World Energy

Needs. Additionally, the theory of technological development is discussed to help readers better understand the nature of the renewable energy market.

The second chapter focuses on the problems, controversies, and solutions for renewable and alternative energy sources and technologies. The sides of the debate over the expiration of fossil fuels are reviewed. Additionally, the chapter addresses the environmental issues associated with continued fossil fuel use and the associated controversy. The estimated prospects for renewable and alternative energy use as the main energy source in the United States are presented. Within this discussion, the advantages and disadvantages of renewable and alternative energy sources are addressed. Furthermore, current U.S. energy policies, particularly those pertinent to renewable and alternative energy, are examined. Finally, policy suggestions for the increased use of renewable and alternative energy are reviewed.

Chapter 3 presents new data about the potential for renewable energy use in various countries. This chapter focuses on the regions of Europe, Asia/Oceania, North and Central America, and South America. Particular attention is given to the European Union, including England, France, and Germany; India, China, Japan, and Australia; and Canada and Brazil. We examine these countries' current policies that encourage the use of renewable and alternative energy. Furthermore, we look at policy suggestions for the increased use of renewable and alternative energy in the international context.

Chapter 4 is a chronology that provides a comprehensive list of the events that have had an impact on world energy issues. Included in the chronology of important energy-related events are technological breakthroughs and scientific discoveries, energy crises, and other political issues. In particular, the focus is on events related to renewable and alternative energy. It is shown that the use of energy allowed humans to create the sophisticated and complex civilization of today. Ironically, energy may end up being the cause of the downfall of the society it helped to build. What is most intriguing is the long history of development of renewable energy and alternative energy technologies alongside the concurrent growth of the fossil fuel industry. Over 100 years ago, engineers and scientists of the day questioned the viability of an economy based on nonrenewable energy and worried that the fossil fuel supply would be exhausted in the near future, causing untold hardships on nations that relied on them as a primary energy source.

Chapter 5 provides the profiles of people who have had an impact on the field of alternative and renewable energy. They have been chosen for their contributions as scientists, inventors, scholars, advocates, and entrepreneurs in the field of alternative and renewable energy. Many of the scientists, inventors, and scholars have laid the groundwork for later developments in alternative and renewable energy technology. Their work is extremely important for making alternative and renewable energy a viable source of power. Others have made the discoveries that highlight the need for diversifying our energy economy, such as the occurrence of global warming and its causes. The advocates and entrepreneurs are responsible for raising awareness of the need for alternative and renewable energy and its possibility for meeting our energy needs. Some disseminate information about the effect of energy use on our environment and health, and they encourage the adoption of responsible policies that take these facts into account.

Chapter 6 is an overview of important facts, data, and documents related to renewable and alternative energy issues. Recent data on energy and renewable energy use throughout the world and in the United States is provided, along with geographical and organizational definitions. An energy conversion table is also provided as a point of reference for comparing various estimates of energy use. The tables include data on the numbers for total world energy consumption, world energy consumption by fuel, and world energy consumption by country. The same is done for the United States, providing an overall look at U.S. renewable energy use and consumption of energy by individual states. Included in this chapter is data on global carbon emissions.

Chapter 7 offers a directory of organizations involved in renewable and alternative energy matters. In this chapter, the organizations' contact information, Web page, and purpose are provided.

Chapter 8 presents print and nonprint resources for readers wishing to investigate the topic further.

A glossary is provided at the end of the book for a clearer understanding of some of the concepts and terms associated with renewable energy.

Zachary A. Smith and Katrina D. Taylor

1

Background and History

Introduction

This chapter provides information on the background and history of renewable and alternative energy issues. It begins by distinguishing between traditional energy sources and renewable and alternative energy sources; included in this first section are definitions of both types of energy. To enhance understanding of energy issues and data, a section on energy conversions has been included. Also provided is a general discussion of the role of energy in society and of the transitions to various energy sources made throughout history. Following this section is the history of renewable energy policy in the United States, which summarizes the federal government's role in the effort to adopt more renewable energy sources. The chapter then describes the dominant energy sources that the United States and the rest of the world currently rely on, followed by a section on the prospect for worldwide energy demand and consumption. The topic of energy conservation is addressed next. A key component for understanding the renewable energy issues is a thorough explanation of the various renewable and alternative energy technologies, which comprises the last section and the bulk of this chapter. Each type of renewable and alternative energy is explained under the following subheadings: History and Uses, Process of Utilization, Advantages, and Prospects for Meeting World Energy Needs. The disadvantages of these technologies are discussed in detail in Chapter 2.

1

Definitions and Distinctions of Energy Sources
Fossil Fuels
Hydrocarbon-based fuels are also known as fossil fuels because they are derived from the centuries-old remains of decayed plants and animals that were captured in geological deposits. These fuels include coal, oil, and natural gas. They are considered nonrenewable energy sources because they take millions of years to form and cannot be easily replenished after use. As fossil fuels became the dominant source of energy, less and less attention was paid to the energy sources of the past: wood, the sun, water, and wind. Developed countries came to rely on fossil fuels for electricity production, for commercial and industrial processes, and eventually for fuel in the transportation sector.

Alternative and Renewable Energy
Renewable energy technologies are those that harness energy from an inexhaustible source. Such sources include the sun, wind, falling water, waves, tides, biomass, or heat generated beneath the surface of the Earth. Most renewable energy sources, except geothermal, are derived from the sun. Solar energy comes directly from the thermal energy given off by the sun. Sources like wind, hydropower (hydro), waves, and tides are the result of solar energy. Wind is created from the differential heating of the Earth's surface by the sun, wind creates waves, and tides are the result of the gravitational pull of the moon and the sun on the Earth. Biomass is actually chemical energy stored in plants that was converted from solar energy through the process of photosynthesis. Renewable energy technologies include solar, wind, hydro, geothermal, biomass, and ocean, which includes tidal and wave power. Many different types of technologies have been fully developed within these categories, and many more are under development. Alternative energy technologies include those that are not derived from fossil fuels but that are also considered nonrenewable. Nuclear power and energy stored in hydrogen are the main alternative energy sources that are examined in this book.

Energy Conversions

To better understand discussions on energy consumption and production, it is important to have an idea of what energy is and how it is measured. Before defining energy physically, two types of energy have to be distinguished: kinetic and potential. Kinetic energy is energy in motion and includes energy types like electrical and thermal (heat). Most renewable energies—like most types of solar, including wind and hydro, and geothermal—fall into this category. Potential energy is energy that is stored in objects and must be released through combustion or chemical conversion. Potential energy can be found in all fossil fuels, biomass, hydrogen, and uranium, the essential chemical for nuclear energy production.

Energy itself can be difficult to define because it has many meanings in different contexts. The widely agreed-on physical definition of energy is the capacity of specific forces to do work. Understanding the elements of work and force in the definition of energy is essential for a full comprehension of energy.

Many forces are capable of doing work, such as electromagnetic and gravitational forces. These forces cause energy to take many forms (thermal, electric, gravitational, etc.). Mathematically, force can be depicted as:

$$Force = Mass \times Acceleration$$

This means that force is the product of an object's mass and acceleration. Force causes an object to alter its motion by changing either its speed or its direction. Force is an essential component in the equation for determining work, which is given by the formula:

$$Work = Force \times Distance$$

The product of this equation is the amount of energy needed to move the object a certain distance. That energy is measured in joules (J). Joules have replaced British thermal units (Btu) as the main measurement for energy. Although these units are the essential units of measurement for energy, other units of measurements are used for specific forms of energy. For instance, thermal (heat) energy can be expressed as calories or watts. Table 6.2 shows conversions for the most common energy measurements and is helpful in gaining a full comprehension of the energy measurements presented throughout the book.

Roles and Transitions of Energy Sources

Human civilization and the use of energy are intricately connected. The advancement of civilization and technology was made possible by the ability of early humans to harness the power of fire and to utilize solar, water, and wind energies without the use of advanced technology. By generating heat and light and transforming it to mechanical and later to electrical energy, hunter-gatherer societies were soon able to shift into agrarian and then into industrial societies. Wind and water were used to power sailing ships and to turn water wheels and windmills for grinding and for other mechanical needs. Fuel wood was the most significant energy source in advanced civilizations like the United States and Europe until the 1800s, and it is still one of the most significant sources of energy in the developing world today. Historically, wood has been used for heating, cooking, and generating steam to run engines until the advent of the Industrial Revolution (Melosi 1985, 19–24).

The intense energy requirements for industrial production of the Industrial Revolution created a need to find a fuel to replace wood. The search led to the discovery of the first hydrocarbon-based fuel: coal. The combustion of coal created the massive energy input needed for the industrial applications that would not have been possible otherwise. Such applications include the production of coke, a coal derivative used for the mass production of iron and steel and for powering steam locomotives, steamboats, and other machinery. Coal also began to replace fuel wood as a source of heat in homes and businesses. Coal maintained its position as the dominant energy source well into the 1800s until oil was discovered in 1859 in Titusville, Pennsylvania. Oil secured its place as the energy source of the 20th century when it was credited for the allied victories in World Wars I and II. The discovery of natural gas soon followed. Energy sources became seemingly abundant, and they could undoubtedly be used for a variety of applications (Melosi 1985, 3). With the discovery of these hydrocarbon-based fuels, industrialized capitalist economies and economies of scale emerged, leading to significant growth in production, population, and wealth. The growth in these latter areas contributed to a stepped-up growth in energy use and consumption.

It has been argued that, for standards of living to increase, the consumption of energy must also increase. Although there appears to be no direct connection between energy consumption and economic output (or quality of life), this point is often debated (Smil 2003, 97–105). There is little doubt among scholars, however, that energy consumption cannot be reduced significantly, much less completely, if the Earth's population continues to increase. Growing populations and development in Third World countries ensure a rising demand for energy. It is also highly unlikely that developed countries will cease energy-intensive operations in the future. While energy-efficient technologies can make the most of current energy usage, human civilization will always require the input of some sort of energy, and overall energy demand will continue to grow even while per-capita energy consumption decreases due to increased efficiency or conservation.

Two major issues are related to these inevitable facts about energy. The first concerns the relationship between energy and the economy. This intricate connection means that energy is necessary to continue economic development and growth. Scarcity of this economic input would create devastating shocks to and possibly the collapse of the world economy. At the same time, the abundant sources of energy, like renewable and alternative, are too expensive for economical development. The initial investment in these energies is high and creates short-term losses for the economy, although the investment pays off in the long run.

The second issue, discussed in the next chapter, has gained significant attention over the last few decades. Current levels of energy consumption have been associated with environmental degradation and the potential for even greater harms to the health of the natural and human world. The question then becomes how can we meet the growing energy needs of an increasing population with the least negative consequences for the environment and the economy? While this book leaves it to the reader to decide the answer to this controversial question, the intention is to provide enough information for someone to come to an intelligent conclusion. To begin answering this question, it is important to understand the different energy sources that are being used and that are available for use, but first we need a description of the history of renewable and alternative energy policies in the United States.

History of Renewable Energy Policy in the United States

The history of energy policy in the United States reflects not only a lack of planning for energy needs on the part of the federal and state governments, but also a perception of energy as plentiful. Thus, long-term planning seemed unnecessary in a time when energy scarcity was a rare phenomenon. Concerns over long-term energy planning were not serious until the last part of the 20th century.

World War I marked the first time that the United States had to address the issue of energy resource scarcity. Federal management of energy production was all-encompassing during both World War I and World War II, though policies and programs guiding these activities were quickly dismantled once the wars were over. The result through much of the 20th century was that actual energy planning was done in a way that Martin V. Melosi refers to as "crisis management," indicating that such planning had only short-term objectives (Melosi 1985, 11). Other energy policies during this era focused more on maintaining a low price of energy supplies for consumers. Through policies that promoted competition, encouraged or discouraged production, and set price controls, the federal government focused more on piecemeal solutions than on a grand plan for a long-term, viable energy economy.

So energy consumption in the United States grew, and eventually its own energy supply became inadequate to meet the needs of its citizens. As a result, it began importing oil, its dominant source of energy, from politically unstable regions like the Middle East. This dependence on foreign oil set the stage for a crisis in 1973. At this time, the Organization of Petroleum Exporting Countries (OPEC) established itself as the dominant world oil cartel after Arab members saw the opportunity to use OPEC's control of oil prices as diplomatic leverage with the West. By collaborating with each other, OPEC members were able to almost double the price of oil. Further adding to the economic challenge that increases in the price of oil caused for the United States, an embargo was placed against the United States by many of the Arab OPEC members in response to U.S. support of Israel in the Arab-Israeli War that was unfolding at the time. While the embargo lasted for no longer than six months, the effects of it are still felt today. During those six months, oil prices in the United States tripled. OPEC established itself as a powerful force with the ability to shake up the world economy for political reasons unrelated to the supply and demand of oil. Factions in the United States have been battling ever since over the need to develop a more reliable energy plan.

The first attempt at creating an encompassing energy policy was President Jimmy Carter's National Energy Plan of 1977. For the first time, the federal government was attempting to aid in the purposeful transition from one energy source to another. Prior to this period, the private market drove energy transitions from wood to coal and then from coal to oil. Economic interests motivated these transitions, not scarcity. The country was not out of wood when it transitioned to coal, nor was it out of coal when oil became the dominant source. More complex economic issues were now driving a government-supported transition from oil and coal to a more stable and sustainable energy source, one that did not have to be imported from abroad. The National Energy Plan focused on reducing oil dependence by reducing energy consumption, improving energy efficiency, increasing utilization of coal by two-thirds, and expediting the use of more solar energy. Within the plan was the Public Utility Regulatory Policies Act (PURPA), which sought to increase renewable energy specifically by creating a market for nonutility electric power producers and by mandating that utility companies buy the power produced by them at a so-called avoided cost rate. Avoided cost is the amount the electric utility would pay to get the energy from another source.

Although Carter's National Energy Plan passed through Congress and the Senate, albeit with significant alterations, it was ultimately abandoned in 1983 when President Ronald Reagan announced his own National Energy Policy Plan. The goal of the National Energy Policy Plan was to maintain energy supply in the most economical way possible, even if that meant importing from OPEC. There was little in the plan to address environmental issues associated with energy use that had begun surfacing in the 1970s. A decline in support for nuclear power and coal because of these environmental concerns conflicted with a desire for cheap and abundant energy to maintain the comfortable American culture and lifestyle. The results of these conflicting values were the continued use of fossil fuels into the 1990s and a slow and ineffective effort to develop sustainable energy.

The two most recent federal efforts to establish a national energy strategy, the Energy Act of 1992 and the Energy Act of 2005, have done nothing to mandate an energy transition and have done little to address rampant concerns about global warming. While many programs and policies are in place to encourage the development of renewable and alternative energy technology, their progress has been slow, and the rate of adoption is not expected to reduce greenhouse gas (ghg) emissions enough to prevent climate change. Although the efforts of the federal government to promote a timely and smooth transition are inadequate, it is helpful to examine the programs and policies that are in place. The specifics of these policies are discussed in Chapter 2. The next section discusses the energy sources currently used throughout the world.

Current Energy Sources

The major source of energy used worldwide is derived from fossil fuels. Oil is presently the main fossil fuel used and hence the most used energy source in the world. The main applications of energy from oil are for transportation, heating, industry, and inputs for industrial production. Energy received from coal is the second most utilized source of energy in the world, mainly for electricity production. Natural gas is the third most used source. Although it is derived from oil deposits, it is not included in data on oil use. Natural gas is used for heating, transportation, and electricity production (Energy Information Administration 2007, 298).

All three of these fossil fuels, naturally stored underground, must be extracted to be used for energy generation, and their extraction methods vary. Oil, most commonly, is extracted by drilling into the earth. Most oil deposits are liquefied and have built up their own pressure, making it easy for them to find their way to the surface once an outlet is created by drilling. In some cases, enhanced recovery is used to create pressure in wells where the oil is too heavy to flow. Enhanced recovery involves drilling a second hole and injecting steam into it to thin the oil and push it through the first hole. Coal deposits, on the other hand, are found only in solid form. The most common extraction method for coal is mining, which can occur in deep underground mines or on the surface in open-pit mines on flat land or mountaintops. Surface mining on mountaintops is known as mountaintop mining and is very controversial. Natural gas, as its name implies, is found in gaseous form. It is extracted in almost exactly the same way as oil: by drilling. Natural gas is then stored in and transported through pipelines.

Oil use accounts for a little more than one-third of total worldwide energy use (Energy Information Administration 2007,

298). The majority of oil use is in the transportation sector in internal combustion engines, the primary technology used for transport, which run on oil or products derived from oil, such as gasoline. Oil is the source for just under 7 percent of worldwide electricity production. Coal is the source of about 25 percent of the energy used worldwide and almost 40 percent of electricity production. Natural gas is the source for about 21 percent of total worldwide energy use and just under 20 percent of electricity use. Nuclear is the next most used form of energy, making up about 6 percent of total worldwide energy used and over 15 percent of electricity used. Finally, renewable energies are the source of 13 percent of total worldwide energy consumption and 2 percent of electricity generation, not including hydropower, which provides over 16 percent of worldwide electricity production (International Energy Agency 2006, 24). Of the various types of renewable energies used worldwide, biomass is used the most, comprising just over 10 percent of total worldwide energy use, leaving hydro to make up 2.2 percent of total energy use and all other renewable energy 0.5 percent of total energy use (International Energy Agency 2006, 6). This amount varies by country. Canada, for instance, derives almost 60 percent of its electricity from hydropower (International Energy Agency 2006, 19). France derives almost 55 percent of its electricity from nuclear power (European Commission 2005, 2.4.1). (See Figure 1.1.)

Projected Outlook for Worldwide Energy Demand and Consumption

In 2006, a study on the world energy outlook by the International Energy Agency (IEA), "an intergovernmental body committed to advancing security of energy supply, economic growth and environmental sustainability through energy policy co-operation," found that, if current trends continue, world primary energy demand will increase as much as 60 percent by 2030 (International Energy Agency 2006, 46). Average annual growth in energy demand is predicted to be less than 2 percent, though in developing countries, especially those in Asia, it is predicted to be much higher (International Energy Agency 2006, 47). The demand for energy in China alone is expected to more than double. In many developing countries, where a fraction of society does not have

FIGURE 1.1

World Energy by Source

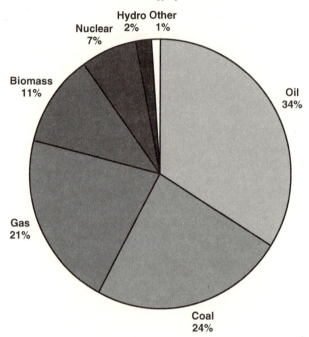

Sources: Energy Information Administration (2007); European Commission (2005); and International Energy Agency (2006).

electricity, energy demand is expected to increase dramatically as these economies and their need for energy and electricity grow (International Energy Agency 2006, 1). Although groups like the IEA and others agree on these world energy outlooks and even though their data is used and trusted by academics and industries alike, the reliability of the forecasts has been questioned. This is discussed in further detail in Chapter 2 in the section on controversies surrounding the oil topping point. The next section addresses the various renewable and alternative energy technologies available for use.

Renewable and Alternative Energy Technologies

The purpose of this section is to provide a thorough review of the various renewable and alternative energy technologies that are at differing stages of development. Before looking at these technologies, however, it is important to understand the widely accepted theory of technology development, which holds that development occurs in stages.

In this context, the various stages of technological development can be related to renewable energy technology. The earliest stage of technology development is known as the conceptualization stage. During this stage, the idea for the technology is conceived. The proof of concept stage follows. In this stage, an idea is taken beyond the conceptualization stage, is converted into an actual model, often on a small scale, and is then tested on a large scale until it is fully operational. The next stage involves tweaking the design until it is ready for market penetration. This stage is where bugs are worked out and the technology is fine-tuned and tested until it has reached optimum performance. Once the technology is ready to meet the user's needs, it can then enter the final stage of market penetration. Market penetration involves the commercial development of a technology that can be used on a mass scale so that it becomes available for widespread usage. While many technologies may be ready for market penetration, several factors can hinder their actual commercial success. These factors include but are not limited to a lack of economic incentive to produce technology on a large scale to be available for commercial use and policy impediments, such as a lack of governmental support for renewable energies and/or the continued support, in the form of subsidies, for fossil fuels, making them more economical than cleaner technologies (Komor 2004, 13).

Many renewable energy technologies have reached the market penetration stage and are ready for widespread usage, but they cannot fully penetrate the market due to several hindering factors, which are discussed in Chapter 2. Other renewable energy technologies are still in the early stages, but they are expected to reach the market penetration stage in the near future. Other alternative energy technologies are in various stages with some, like nuclear fission, already commercially used on a mass scale. Regardless of a technology's stage, the future potential of all of them to contribute to the energy supply is great.

Solar

History and Uses

Solar energy is simply the capture of heat energy from sunshine. Solar energy has been used domestically for heating and cooking since the dawn of human civilization. Ancient societies used methods of passive solar capture to heat dwellings. It is therefore difficult to pinpoint exactly when it was first used. The first modern solar technology can be traced back to the late 1700s when Antoine Lavoisier, a Frenchman, developed the first solar furnace. Solar was used in the 1880s in India for cooking. Simple steam engines have been powered by solar since 1878. In 1913, Egypt developed a solar-powered water pump for irrigation. And solar hot water heaters have been popular in the United States since the 1930s. The first photovoltaic (PV) cells, used to produce electricity, were developed in the late 1950s and were used to provide electrical power for satellites orbiting the Earth. Improvements to PV technologies in the 1970s helped to reduce their costs and allowed for their penetration into more applications, all of which met low-power needs. After the energy crisis in of the 1970s, PV power systems with utility grid–connected applications emerged worldwide, and solar power began to be seen as a viable energy source.

Solar energy is converted to thermal energy or electrical energy. People and countries around the world heat their homes and water, cook, and generate electricity using solar energy. At the start of the 21st century, the four most important markets for solar energy use were communication industries, recreational uses, home solar systems (including the generation of power in villages in low-income countries), and water pumping (European Commission 2001, 7).

Process of Utilization

Solar power is utilized through one of three methods. The first is solar thermal systems in buildings, which capture solar energy in two ways: actively or passively. For passive solar heating, buildings are designed in a way that captures heat and transmits daylight. Basically, this method requires the strategic placement of windows in building design so that the maximum solar energy potential is captured. Active solar heating is essentially the capture of solar energy using collectors in which the thermal energy is usually transmitted through a liquid medium. This type of solar energy is generally used for hot water heaters and for space heating and cooling (Tester et al. 2005, 557–559).

The second method of utilization is through photovoltaics. Photovoltaics are solid-state semiconductor devices that convert sunlight directly into electricity, which can then be used for several types of applications. Photovoltaics require no moving parts or fluids, and they are made of one of three types of silicon or come in the form of a thin film made from other semiconductor materials. PV systems can be stand-alone, roof mounted, or developed as roof shingles. The world's largest photovoltaic solar power plant is in Pocking, Germany, which produces 10 megawatts at its peak. An even larger plant is being constructed in Portugal and is expected to produce 11 megawatts at its peak (PowerLight 2006).

The third method of utilization is a technology called a central solar power (CSP) system, also known as a concentrating system. Sunlight is focused or concentrated to generate steam, which is then used to run turbogenerators. There are three types of CSP systems. (1) Parabolic trough systems use concentrators that focus sunlight into a tube that runs the length of the trough. (2) Towers collect sunlight using fields of sun-tracking mirrors known as heliostats and concentrate the heat onto a centrally located tower-mounted receiver where it heats a liquid, which is constantly circulated. (3) Parabolic dishes, or dish-engine systems, concentrate sunlight onto a focal point in the middle of a dish, where the heat is then transferred to an engine and converted to mechanical energy. All of these technologies can be used to heat and cool air or water and to generate electricity (National Renewable Energy Laboratory 2001, 1–3).

All of these technologies have reached the final stages of development and are ready for market penetration. Many other solar technologies have yet to make it out of the proof of concept stage. Lunar solar power (LSP) systems, which would capture solar energy on the moon and transmit it to the Earth via microwaves, are currently being developed (Criswell 1997, 24–26). Also on the drawing board are space-based solar power satellites that would capture solar energy while in orbit around the Earth and transmit it through the atmosphere to the ground as microwaves; there the microwaves could be converted into energy.

Solar sails use sunlight as wind is used on Earth to power vessels using large lightweight mirrors as sails (McInnes 1999, 17). Given current economic and policy trends, such technologies are decades from reaching the final stage of market penetration.

Advantages of Solar Power

Experts and environmentalists alike have touted the many advantages of solar power. One of the main advantages is that it is an inexhaustible resource, which, of course, is true of most renewable energy sources. More importantly for the case of global warming and air pollution, it is a comparatively clean renewable energy and has very low ghg emissions. Environmental impacts, like pollution, would be negligible except during the manufacture and transport of the technology. The development of thin-film photovoltaics seeks to remedy this problem by requiring even less energy and toxic materials in the manufacturing process. The technology is flexible and can be used for a broad array of applications. Like many other renewable energies, it has a modular characteristic; that is, the site of development does not have to be permanent, reducing the likelihood that the technology will be locked in geographically. Furthermore, large-scale solar power technologies require much less land than other energy sources, such as large-scale hydro (dams), biomass, and coal technologies to produce the same amount of energy.

Solar Potential to Meet World Energy Demands

Worldwide solar energy potential is several times the amount of global energy consumed annually by humans (World Energy Council 2001), and there is little doubt that it could be used at least to offset dependence on fossil fuels. CSP systems have especially high potential for energy production in arid, desert regions like the American Southwest with the possibility of transmitting excess energy to other regions. Scientific calculations have been made indicating that installing CSP technology on less than 1 percent of the world's hottest deserts could provide for the world's electricity needs (Seager 2006). The integration of photovoltaic technology in new building designs can lessen the need for land while increasing the potential for small-scale energy production and fulfillment. When integrated into building structures as such, a solar power system has the potential to provide for all the energy that building needs to operate. Although photovoltaics are still relatively inefficient in converting energy compared to other technologies, large gains in efficiency are achieved every year as the technology develops (Green et al. 2004, 369). The solar potential for the transportation industry rests on the development of technology to store energy derived from solar power, an issue that is discussed further in Chapter 2.

Wind
History and Uses
Wind is the result of the differential heating of the Earth's surface. Differential heat is the difference between how much heat is absorbed by land and how much is absorbed by water surfaces. Wind energy is captured via a turbine, which turns as the wind blows. The turbine generates mechanical or electrical energy, which can then be used for a variety of applications. Wind was used for centuries to power ship sails. The earliest documentation of windmills can be found in the writings of an Islamist geographer who recorded the use of vertical axis windmills in Persia in 9 CE (Hassan and Hill 1986, 54). Windmills have been used in Europe since the 12th century, and their use has been documented in China since the 13th century. The first commercial wind turbine that produced electricity was built in Vermont in 1939. Currently, the primary use for wind power is electricity generation.

Process of Utilization
Unlike solar power, wind power is captured with essentially only one type of technology: turbines. Large-scale turbines are made up of two rotors consisting of two or three propeller-like blades attached to a nacelle (the housing containing the electric generator, gear box, and brake), which is held up by a support tower. Small-scale turbines are similar to the large-scale types, but the rotor powers the electric generator directly without a gear box, and they tend to be only 25 to 30 feet high. Horizontal axis wind turbines (HAWT) are much more common than vertical axis wind turbines (VAWT). HAWTs are similar to propellers, whereas VAWTs look exactly like eggbeaters. Also, VAWTs tend not to be as efficient as HAWT units, which is why all grid-connected commercial wind turbines are horizontal. Wind turbines are usually grouped together within the same area of land. These groupings, called wind farms, generate bulk electric power

to be fed into the grid like traditional forms of energy (Herbert et al. 2007, 1117).

Advantages of Wind Power

The benefits of wind are arguably greater than those of any other renewable energy technology. Like solar, wind is virtually inexhaustible, and it produces almost no ghg emissions. In fact, it is touted as having the lowest environmental impact of any renewable energy technology on the market (Herbert et al. 2007, 1118). The production of wind technology is low energy–intensive. It is the most economically viable of all renewable energy technologies with the most potential to compete in a fossil fuel–dominated market.

Prospects for Meeting World Energy Needs

If wind energy technology continues at its current rate of adoption, it will likely meet 10 to 20 percent of the world's electrical needs by 2050. Exploiting the wind potential in North and South Dakota alone would provide for almost half of the current electricity needs of the United States. Worldwide offshore wind potential is large enough to provide for the current electricity needs of the entire planet. Recent studies show that, if even 20 percent of potential wind energy sites were exploited, they could produce enough energy for the world's needs (Archer and Jacobson 2005).

Biomass

History and Current Uses

Biomass consists of virtually any living plant material as well as organic wastes from sources such as humans and plants. The storage of solar power through photosynthesis creates products with a high-energy content, like carbohydrates, which can be used as an energy source when combusted. The ability to generate large quantities of biomass on a short timescale makes it possible for continued human use of them. Hence, they are considered a renewable energy.

According to Mehrdad Arshadi et al. (2004, 108), biomass takes one of four different forms:

1. *Biogas:* Methane is the most commonly used biogas and is emitted through the decomposition of organic waste.

2. *Solid biofuels:* These can be derived from agricultural crops, forests, organic waste, and hay.
3. *Liquid biofuels:* Biodiesel and ethanol are the most common liquid biofuels and have received a considerable amount of attention in the media over the last decade as the political windows have opened for favorable legislation.
4. *Energy crops:* These are crops grown specifically for use as a solid or liquid biofuel. Corn and sugar cane are the most common crops, grown for the production of ethanol.

Historically, biomass was the primary source of fuel for the world until the Industrial Revolution of the 1800s brought about the transition from wood to coal. Biomass is the most widely used renewable energy in the world due to the fact that many developing countries rely on it for heating and cooking. Even more complex applications are being used in developing countries. For example, Brazil has taken the lead in converting its transportation sector to use ethanol as a primary fuel in place of gasoline. Also, biogas producers, which capture methane from the decomposition of municipal waste in landfills, are used extensively in China and India. Due to its extremely versatile nature, biomass can be used in any application that fossil fuels are used in. It can be used to generate electricity, for commercial and residential heating, and for transportation. It has surpassed hydropower as the most used renewable energy source in the United States (Energy Information Administration 2007).

Due to the growing importance of biofuels on the political agenda of countries throughout the world, different types of biofuels are being promoted. In the United States, ethanol from corn is the focus of legislation to promote this biomass as a fuel source. Although corn ethanol can be grown in the United States, thus aiding in achieving the goal of energy independence, the concern is that the use of corn for this purpose could lead to significant soil degradation and pollution from the use of fertilizers needed to grow it and that it could encourage the use of genetically modified varieties of corn. There are also concerns that using corn for fuel will cause the price of corn to increase, resulting in heightened food prices because corn is used to make many types of food and to feed livestock that is subsequently eaten. In Brazil, sugarcane is the dominant source for ethanol. Sugarcane ethanol is widely thought to be the most efficient of biofuel sources, and the cane can be grown on marginal soils, making it less likely to displace land needed for agricultural production. Some are concerned over the environmental consequences associated with the pollution emitted from sugarcane ethanol. In Europe, the dominant biofuel is biodiesel made from soybeans. Soybean biodiesel emits much fewer pollutants than diesel made from petroleum, making it a more environmentally sustainable fuel. There are concerns, however, that clearing the land needed to grow the soybeans would result in deforestation and that this could have a negative effect on the reduction of greenhouse gases and on biodiversity. The burning of biodiesel still produces air pollution, though about half that of petroleum-based diesel fuels. Biodiesel can also be made from cooking grease. This source of biofuel is attractive because it recycles a material that would otherwise be wasted. Like soybean diesel, it still causes air pollution and the emission of greenhouse gases. Although it is not being commercially produced yet, cellulosic ethanol derived from switchgrass, slash, and agricultural byproducts has been getting attention in scientific circles. This is because cellulosic ethanol has the real potential to significantly reduce greenhouse gases. There are concerns about this source, however, because widespread growth of switchgrass could displace native plants and wildlife habitat.

Process of Utilization
Biomass can be converted to energy through a number of processes. The main method of utilization is through thermal conversion, which is essentially the combustion of biomass to create energy or fuel. The process of bioconversion is the direct or indirect use of the chemical composition of living things to transform one substance to another. Fermentation and hydrolysis are used to create ethyl alcohol or ethanol, a high-quality fuel that can be used for transportation or electricity production. Biogas producers, already mentioned and also known as digesters, are contraptions that use anaerobic bacteria to process animal waste and sewage as well as landfill material to create methane, which is then captured and used.

Biomass has a wider range of potential uses than wind and solar energy. Like the latter two, it can be converted to electricity. It can also be burned with coal to reduce sulfur emissions and to help with the transition from fossil fuels to renewable energies. It can be used by itself to generate electricity by being converted to fuel, which can then be combusted to make steam or power a gas turbine or used in fuel cells for direct conversion to electricity. A variety of biomass fuels, in the form of synthetic gases like ethanol, biodiesel, a combination of alcohol and vegetable oil, animal fat or recycled cooking oils, can also be used to power the transportation sector, which is currently a significant source of greenhouse gases.

Advantages of Biomass

Biomass has many of the same benefits as other renewable energies. It has high availability across the globe, unlike fossil fuels. Because biomass consists of plant matter and refuse, it can be found in virtually every country in the world rather than being limited to a few specific parts of it. It has the characteristic of being renewable because, when biomass crops are grown sustainably, they become an inexhaustible source of energy. Although the combustion of biomass can emit carbon dioxide, it also acts as a carbon sink when it is grown. As long as it does not emit more carbon dioxide when used than it takes in while growing, it essentially becomes a zero net emitter. Studies have shown that ethanol, when burned as a fuel, emits as much as 38 percent less greenhouse gases than gasoline (Shapouri, Wang, and Duffield 2006, 85). Cellulosic ethanol, as opposed to ethanol made from corn, has the potential to offset even more (Hammerschlag 2006, 1749). It emits much less sulfur dioxide than fossil fuels, so it would cause less acid rain. Biomass has the added benefit of natural storage, making it available at all times without the fluctuations that can occur with wind and solar. It also provides an alternative for managing municipal and agricultural waste, the former of which has become quite a problem over the last few decades. Because it can be cofired with fossil fuels, it has been cited by many experts as the renewable energy that can help phase out fossil fuel use and smooth over the transition to cleaner energies like wind and solar.

Prospects for Meeting World Energy Needs

Worldwide production of energy from biomass has the potential to provide for over half of the world's energy needs by 2050. Biomass and particularly ethanol have a high potential for offsetting the need for fossil fuel–dominated energy sources. This is particularly true in the transportation sector. For electricity production, however, biomass may be best suited for areas that are not highly populated and do not require high-intensity use of energy. The potential for high-intensity use is much lower than other fossil fuels because of the amount of land required to grow biomass crops. It could be especially useful in handling municipal waste, although it is uncertain how much energy could actually be provided by waste in highly populated areas.

Geothermal

History and Current Uses

Geothermal energy is heat that is stored in the Earth's crust, usually beneath the ground. It is created by the decay of radioactive minerals underneath the Earth's surface. Hot springs and geysers are examples of geothermal energy that has traveled aboveground through the medium of water or steam. This type of geothermal energy source is known collectively as hydrothermal or convection-dominated, referring to the production of hot fluid. This category also includes boiling mud pots and fumaroles (steam or gas vents). Hydrothermal sources are generally found in mountainous areas, near volcanoes, and at the edges of crustal plates. Distribution of these sources is largely related to plate tectonics. Steam-dominated systems are the least common of all the geothermal sources, and they are the primary sources and locations for geothermal power plants because their temperature can get high enough to be converted into electricity. Geothermal energy can also be found in hot dry rock. Hot dry rock is distinguishable from hydrothermal due to the lack of fluid. This is the most common form of geothermal energy systems and also more difficult to extract heat from than hydrothermal systems. Wells must be drilled into the rock and fluids circulated to transfer heat from underground to where it is needed aboveground. The last type of geothermal system is magma. Magma is rock that is partially or completely molten. It can usually be found only at great depths, requiring intensive drilling and making extraction the most difficult, although not impossible, of all the geothermal resources.

Geothermal energy has been used and is still used for cooking and bathing. The earliest usage probably dates back to prehistoric times when humans likely bathed in hot springs. It was first used for commercial electricity production in the early 1900s in Larderello, Italy, from steam vents and drilled holes in what is known as the Larderello field. Use of geothermal for electricity generation has expanded to the United States, Japan, Mexico, Iceland, and the Philippines, to name a few. Geothermal ranks third in the United States for renewable energy use behind biomass and hydropower (Energy Information Administration 2007).

Process of Utilization

To utilize geothermal systems, access to the sources of heat must be gained. This generally involves drilling techniques similar to those used by the oil and gas industry. Liquid is essential because it is the medium needed to transport the heat. It must therefore either exist at the source, as in hydrothermal systems, or be brought in externally and run through pipes where it can be heated by hot dry rocks or magma and circulated to generate energy. The use of geothermal is very similar to that of solar. It can be incorporated into building structures through the use of pipe systems for heating and cooling when the structures are near the source. It can also be used to generate electricity in large-scale power plants, which must also be located near the source. It has found a niche in the agriculture industry where it has been used to heat soil and greenhouses and to dehydrate foods.

Advantages of Geothermal Energy

Geothermal is a clean energy source releasing practically zero emissions (Energy and Geoscience Institute 2001, 7). Its use causes a negligible amount of pollution (sometimes unpleasant although generally nontoxic smells). Geothermal power plants can be designed to emit no pollution at all (Bjornsson et al. 1998, 4). It has a large worldwide resource base and can be found in many countries. Unlike other renewable energies, it is not subject to fluctuations in supply and is constantly available.

Prospects for Meeting World Energy Needs

Geothermal energies, if captured, have a higher energy content than other renewable energy sources. While the resource base is not yet fully known, it is estimated that reserves could contribute a significant amount of energy to the growing global consumption. It is estimated that geothermal could help offset global warming emissions by as much as 10 percent by 2020 with just the present growth rates (Bjornsson et al. 1998, 5). Technology in this field has the advantage of being shared with the lucrative fossil fuel industries since the method of extraction (drilling) is the same. Much of the research and development from these industries can be used to improve geothermal development.

Hydro

History and Current Uses

Hydropower is energy captured from falling or running water, usually rivers. Hydropower technology can be developed on a large or small scale. Large-scale hydro generally refers to large dams. The specific categorization is contingent on the amount of electricity produced by a hydropower project. Small-scale hydro captures energy through the use of smaller dams and river turbines. It is usually considered small scale if it produces fewer than 50 megawatts of electricity, but the definition varies by country.

Historically, hydropower has been used to run mills and to generate mechanical energy for industrial uses. It was first used to generate electricity in 1881 in Surrey, England. Currently, there are over 45,000 large dams in over 140 countries, although not all of them generate electricity (World Commission on Dams 2000). Alternative uses for dams include the storage of water in reservoirs for irrigation and other water needs, as well as water-oriented recreation.

Process of Utilization

The most widely adopted process of utilizing hydropower is through large-scale hydro projects that involve the construction of an impoundment dam on a river. The dam holds water in a reservoir and directs the flow of water into a penstock, which carries it to the turbines installed in the dam. The turbines are connected to electric generators, and, as the water spins the turbines, the generators create electricity. When electricity does not need to be generated, water is kept in storage in the reservoir for future use. Small-scale projects are generally referred to as micro hydropower. They include projects known as "run of the river" that use the natural flow of the water to spin turbines or waterwheels. They do not interfere with the flow of the river and do not include structures, like dams, that store water in reservoirs.

Advantages of Hydropower

The advantages for hydropower are distinct for large- and small-scale projects. Large-scale projects have low operating costs and longer expected power plant life than any other form of electricity production. This fact makes for very low consumer prices. Reservoirs often serve multiple purposes, including providing a source for irrigation, drinking water, recreation, flood protection, and resources for aquaculture. Environmentally, they emit almost no carbon dioxide or particulates compared to other fossil fuel sources of electricity.

Small-scale hydropower is particularly useful for low energy–intensity uses like those of developing countries. It is more economically feasible for countries that do not have the capital to invest in large-scale hydropower. It is also beneficial for remote areas where the transmission of electricity is difficult. They have all of the benefits of large-scale hydro, except for those related to the reservoirs themselves like recreation, and hardly any of the disadvantages. Therefore, they have minimal environmental impacts compared to large-scale projects and fossil fuel usage.

Potential to Meet World Energy Needs

Much of the potential for hydropower has been tapped in areas where it has been intensely developed, as in the United States and most of Europe. The potential for the capture of hydropower on a global scale, however, is still very high. "Two-thirds of 'economically viable' hydropower potential is yet to be tapped" (Beijing Declaration on Hydropower and Sustainable Development 2004). The vast majority of this potential exists in developing countries, many of which have thousands and even millions of citizens without access to electricity. Unfortunately, the areas where hydropower could be developed tend to be densely populated. Developing large-scale hydropower in these areas would cause the displacement of millions of people. The worldwide potential is unevenly distributed across continents, making hydro a useful technology for some countries but not for others. For example, Asia holds 46 percent of the world's hydropower potential, with China making up an additional 15 percent of worldwide hydropower potential (World Bank 2007).

Ocean Energy: Wave, Tidal, and Ocean Thermal Energy Conversion

History and Current Uses

Ocean energy is caused by the gravitational pull of the moon, sun, and Earth (creating tides), by wind (creating waves), and by temperature differences in the water. Medieval times saw a fairly widespread use of ocean waterwheels across Europe. While many processes for the utilization of ocean energy have been considered, only a few methods have received the most research and development attention. The three most developed types of ocean energy technologies are wave, tidal, and ocean thermal energy conversion (OTEC).

Process of Utilization

Many technologies are currently being developed to capture wave energy. These technologies extract energy from surface waves or from pressure fluctuations beneath the water. Wave energy can be converted to electricity in both onshore and offshore applications. Offshore applications are located in water that is over 100 feet deep. The motion of the waves is used either to power pumps that create electricity or to pressurize water that then turns a turbine. Special sea vessels have been developed that can generate electricity from waves as they travel across the ocean by funneling water inside the craft to turn turbines and then releasing it back into the ocean. Onshore applications make use of the power of breaking waves on the shore. Three technologies have been developed to utilize this energy source. The first is the oscillating water column. Through the use of a concrete column located in the water, air is compressed and depressurized by water that enters the column from wave motion. It is then used to spin a turbine, which then produces electricity. The second technology is a tapchon, or tapered channel system, which feeds into a reservoir located on a cliff above sea level. The narrowed channel causes the waves flowing inside of it to increase in height and spill over into the reservoir, where the water is used to run turbines in a manner similar to that of large-scale hydropower applications. The final device used to convert wave power into energy is a pendulor device, which is a rectangular box that is open to the sea at one end with a flap hinged over the opening. As the flap is moved back and forth by the wave motion, it powers a hydraulic pump and generator. Although few of these technologies have reached the market penetration stage, the first commercial wave power station came online off the island of Islay off the coast of Scotland in 2000.

Tidal energy technologies utilize ocean currents and the difference between low and high tide to generate electricity. The difference must be great enough, over 16 feet, to create electricity. As in wave power, three main technologies have been developed to convert tidal power into electricity. The first, a barrage or dam, forces tides into an area where they come into contact with a turbine and power a generator. The second technology is known as tidal fences, which are apparatuses that look like giant turnstiles. The turnstiles, spun by ocean currents, are capable of generating electricity at rates as high as winds with very high velocity. The last technology is the tidal turbine. Similar to wind turbines or river turbines, tidal turbines are situated underwater in an arrangement that is similar to that of a wind farm. At about 50 feet in diameter, they are ideal in waters that range from over 60 to 90 feet in depth. The only large-scale operation using tidal energy is in La Rance, France.

Ocean thermal energy conversion is a technology that is still in the developmental stages and that is intended to extract power from temperature differences between the surface of the ocean and deeper waters through the use of a heat engine. Georges Claude, a French engineer, built the first OTEC power plant off the coast of Cuba in the 1930s. Claude built another plant off the coast of Brazil. Both plants were destroyed by wind and waves before they were able to produce more power than they needed to operate.

There are three types of OTEC technologies. The first is called a closed-cycle system. Low-boiling point fluid is vaporized by warm surface water that is pumped into a heat exchanger. The vapor is used to rotate a turbine, which generates electricity. Cold deep seawater is then used to condense the vapor back into a liquid where it is recycled into the system. The second type of technology is the open-cycle system. When warm surface seawater is placed into a low-pressure container, it boils, producing steam that can turn a turbine and generate electricity. In this system, the steam is turned into freshwater and exported from the system, and new ocean water is used as input for further energy production. The final type of technology is the hybrid system. This system uses the steam from the open-cycle process of boiling water to vaporize the low-boiling-point fluid used in the closed-cycle system, which is then used to turn a turbine and generate electricity.

Advantages of Ocean Energy

Despite the need for further development, the potential exists for many benefits if ocean energy can be efficiently captured and utilized. Small island societies stand to gain the most from small-scale use due to their low energy intensity. Ocean energies are more predictable and reliable than other renewable energy sources. These technologies could be paired with offshore wind operations, thus maximizing the use of such spaces. The ratio of the energy needed to make the technology to the energy created by it is low compared to other energy sources. Emissions would be nil or, in the case of OTEC, much lower than fossil fuels. Any negative environmental impacts from these technologies are reversible and not global in nature. OTEC, in particular, has very high energy potential despite its low energy conversion efficiency. Another important and interesting advantage of OTEC is that the process produces not only energy, but also water that is desalinized enough for human consumption.

Potential for Meeting World Energy Needs

If developed, ocean energies have a greater potential for meeting world energy needs than any other source. The potential of all ocean energy technologies is so high that harnessing just a fraction of a percent of the world's wave power could provide five times the global demand for energy. There are 40 sites in the world where tidal power could be converted to electricity. Where it could be utilized, ocean energy could provide for one-third or more of that area's electricity needs. OTEC also has the potential to provide for all of the world's energy needs if storage and economic obstacles can be overcome (Avery and Wu 1994, 3).

Alternative Energy

Nuclear Fission

History and Current Uses

When the first nuclear chain reaction was achieved in 1942 at the University of Chicago, the possibility for energy production from this source seemed infinite. The first commercial nuclear reactor was built not long after in 1956 in Calder Hall, Britain. It would become the first of hundreds of nuclear power plants. The global total of operating plants today is 444 in 25 countries (American Nuclear Society 2006, 37). Although nuclear power provides a considerable amount of the world's energy (over 6 percent) and electricity (17 percent), compared to renewable energy, the rate of adding nuclear power reactors to the grid has been slow globally (less than 1 percent annually). Only a handful of reactors have been added to the grid in the United States since 1973, and they have come online only because they had been licensed prior to the moratorium on licensing new reactors. The slow growth is due to an aggressive and successful grassroots campaign against the use of nuclear power because of safety and environmental risks. In 1979, those concerns were justified as a nuclear reactor meltdown occurred at the Three Mile Island plant in Pennsylvania. Just seven years later, the world was horrified as it watched the aftermath of a nuclear explosion at the Chernobyl nuclear power plant in Russia, which led to the release of dangerous radiation. Despite these accidents, the development of nuclear power continues to occur globally, albeit cautiously, to meet the rising world energy needs.

Process of Utilization

Nuclear power requires the input of nuclear fuels like uranium or thorium. These elements are mined out of the Earth using the same techniques as coal mining. Ninety percent of the world's uranium is mined in Australia or Canada. In the process of nuclear fission, nuclear fuels are placed into a nuclear reactor core, where a self-sustaining process of atom splitting occurs and generates heat. The heat is then used to create steam, which then powers a turbine to generate electricity. Once the reaction no longer produces enough heat to create steam, the reactor cores must be refueled for the process to start again. Six types of reactors are used worldwide, but they all utilize nuclear fission to create energy. The pressurized water reactor (PWR) is the most commonly used. There are a total of 268 online with the majority in the United States, Russia, Japan, and France. It uses enriched uranium as its fuel. The second most common reactor is the boiling water reactor (BWR). There are a total of 94 worldwide, mostly located in the United States, Japan, and Sweden. Like the PWR, the BWR uses enriched uranium as its fuel. The pressurized heavy water reactor (PHWR), also known as the CANDU, is exclusive to Canada, with a total of 40 plants in operation. It uses natural uranium as a fuel and pressurized water as a coolant. The next most common reactor is the gas cooled reactor. A total of 23 exist and are found only in the United Kingdom. It requires natural and enriched uranium and uses carbon dioxide instead of water as a coolant. The light water graphite reactor (RBMK) is found only in Russia. There are a total of 12, and they rely on enriched uranium for fuel. The last type of reactor is the fast neutron reactor, also known as fast breeder reactor (FBR). There are only four plants online, located in Japan, Russia, and France, and only one is used commercially. Because they use plutonium as well as uranium to fuel the reactions, they are capable of getting 60 times more energy from uranium than other reactors.

Advantages of Nuclear Power

Compared to fossil fuels, nuclear power is a much cleaner source of energy. Nuclear fission creates zero ghg emissions, making it a better alternative than fossil fuels in light of global warming risks. It also emits none of the gases related to acid deposition, also known as acid rain. Because uranium can be transported, nuclear power plants can be built anywhere, making them a very modular energy source. It is possible therefore to build them in areas of high energy usage, mitigating transmission problems. The production of electricity from nuclear fission is fairly efficient compared to renewable energies.

Prospects of Meeting World Energy Needs

Nuclear power has a high potential for meeting world energy needs. Many experts contend that nuclear power is the solution to becoming independent of fossil fuels for energy production. Although the world's electricity needs could be met if enough nuclear power plants were constructed, there is no guarantee that the plants would have enough uranium. Studies indicate that uranium shortages could begin within the next decade, and the problem would only become more acute if demand for nuclear power increases. Because uranium is essential for nuclear power production, a shortage would make nuclear energy a nonviable option for replacing fossil fuels as our main energy source.

Fusion

History and Current Uses

Whereas in nuclear fission the atoms of heavy elements are split to create energy, in nuclear fusion light elements are combined, or fused, to create heavy elements. Nuclear fusion is the dominant reaction in thermonuclear weapons, but it has not been used to generate energy. As of 2007, nuclear fusion has not passed the proof of concept stage in its technological development, and it is not likely to do so for decades to come. International fusion research programs, however, are currently underway to discover how to harness this energy source. In 2006, the United States, the

European Union, China, India, Japan, South Korea, and Russia signed an international fusion energy agreement to build an international fusion project called ITER. It is being built in France, and completion is expected in 2015. The ITER project itself will not be used to produce electricity from nuclear fusion on a large scale, but it will be used to generate the data needed to move forward in the hopes of building the first nuclear fusion power plant. Although nuclear fusion has been so far unable to provide any energy for human needs, it is a technology with great potential.

Process of Utilization

The most promising type of nuclear fusion requires the light elements deuterium and tritium. The combination of these elements creates helium and considerable amounts of energy. Currently, no successful process of utilization on a grand scale has been discovered. For fusion of deuterium and tritium to occur, they must be heated to 100 million Kelvin, where they take the form of plasma, and scientists are trying to learn how to control high-temperature plasma. The core of the sun is 15,000 Kelvin, making the deuterium and tritium plasma an extremely hot form of matter, and the material needed to store matter at such a hot temperature has yet to be discovered. The challenge for scientists and engineers is, then, heating these elements to 100 million Kelvin, containing them, and sustaining a stable fusion reaction. Cold fusion, also known as low-energy nuclear reactions (LENR), has yet to be realized. In this theoretical type of fusion reaction, the elements do not need to be heated to such extreme temperatures, alleviating many of the challenges that scientists are presently trying to overcome. Although many scientists are working to prove that cold fusion is possible, the top countries involved in the initial research, including the United States and Japan, have abandoned their efforts. Although whether such a process is possible is unknown, there is little hope in the scientific community that cold fusion will ever be realized and able to contribute to large-scale energy generation.

Advantages of Nuclear Fusion

Nuclear fusion enjoys many advantages over nuclear fission. It is much more environmentally friendly. Not only does it emit no greenhouse gases or acid deposition compounds, it does not produce radioactive waste that needs to be disposed of and stored. Furthermore, the element needed to fuel fusion reactions, deuterium, can be economically obtained from the ocean and is therefore virtually inexhaustible. Although tritium does not occur naturally, it can be created from lithium, which is also very abundant. Nuclear fusion has an extremely high energy density compared to renewable energies. Although the successful utilization of nuclear fusion is not expected to happen anytime sooner than 50 years from now, medical, civilian commercial market, environmental, and defense commercial opportunities exist to help fund the cost of fusion research in the short term.

Prospects for Meeting World Energy Needs

Perhaps the best argument for continuing research into nuclear fusion is its potential for energy production once a reaction can be contained and sustained. Fusion reactions release extremely high amounts of energy. If that energy can be harnessed and transformed into electricity, fusion could provide a significant amount of power and contribute to a reduction in reliance on fossil fuels (Tokimatsu et al. 2003, 5).

Energy Conservation

While energy conservation is not a source of energy, it can offset enough energy use that it could be considered as an alternative to the continued or increased use of fossil fuels as an energy source. The United States currently uses over 97 trillion joules annually (British Petroleum 2007, 37). Of that amount, it is estimated that as much as 32 trillion joules could be saved annually in the United States if energy conservation strategies and alternate technologies were implemented (Pimentel et al. 2004, 281). Furthermore, the same study indicates that the United States would be able to completely eliminate dependence on imported oil through energy conservation efforts.

Energy conservation and efficiency techniques in the transportation sector include increased vehicle fuel economy (better gas mileage); increased carpooling and increased use of public transportation; shifting to trains for commercial shipping; and incorporating energy efficiency innovations into current shipping transport methods like trucks, planes, and ships. In the building sector, energy wasted through heating and cooling could be drastically reduced by better sealing of air leaks, better insulation, the conversion of existing windows and the original construction of new windows to improved window design, the incorporation of passive solar design, and the use of high-efficiency furnaces and air conditioners. Increasing the energy standards for equipment and appliances like washing machines and refrigerators and the use of energy-saving software would also reduce energy consumption. Lighting represents a large proportion of energy use. The conversion of current lighting to improved efficiency lighting, such as compact fluorescent lights, could drastically reduce energy consumption. Chapter 2 provides a discussion of dissenting opinions on the effectiveness of energy conservation and efficiency for reducing energy consumption.

Hydrogen

Hydrogen, discussed briefly here, is a more appropriate topic for Chapter 2. The use of hydrogen for energy purposes is a fairly new endeavor, although its possibility has been discussed and researched for well over a century (Hoffman 2001, 28). Technically, hydrogen is not a source of energy, but rather a carrier of energy. Hydrogen cannot be mined or collected using technology; it must be created, like electricity, with either fossil fuel or renewable energy sources. It is created through a highly energy-intensive process called electrolysis, which is the use of electricity to split chemical elements and compounds. In the case of hydrogen, the water compound (H_2O) is split to separate the hydrogen element from the water molecule, leaving oxygen as the by-product. Most electrolysis today is powered by fossil fuels, making the creation of hydrogen unsustainable. Most experts predict that in the future hydrogen will be created using renewable and alternative energy sources like those discussed in this chapter (Dunn 2001, 23). Hydrogen is mainly used to make ammonia and methanol, to refine oil, to hydrogenate food oils, and as a rocket propellant. When it is converted into a fuel like gasoline, it can power micro turbines to create electricity and power vehicles. Used in a fuel cell, it mixes with oxygen to create electricity and emits only water. The potential of hydrogen to meet world energy needs without creating severe environmental degradation is heightened when it is created using renewable energy.

Conclusion

Because energy consumption is unlikely to decrease in the future, there is no doubt that energy issues will remain at the top of political agendas around the globe. This chapter has illustrated the uses and sources of energy throughout the world. Fossil fuels remain the dominant source of energy by far, but renewable energy technologies are growing by type and viability. Even though energy conservation can result in large overall energy savings, energy production to meet the growing needs of the world is still necessary. Renewable energy sources such as solar, wind, biomass, geothermal, hydro, and ocean are ready for use on a mass scale. Alternative energies like nuclear and hydrogen are also being used and developed for widespread usage. This chapter has examined each of these energies, their history and current use, their process of utilization, their advantages, and their prospects for meeting world energy needs. The next chapter presents an in-depth look at the problems and controversies surrounding the development of nonfossil fuel energy and the debate over the need for a transition away from fossil fuels toward renewable and alternative energy sources. The main focus of the chapter is on the United States, and it includes a full survey of policies in place to encourage renewable energy production and use.

References

American Nuclear Society. *Nuclear News World List of Nuclear Power Plants.* La Grange Park, IL: Author, 2006.

Archer, Cristina L., and Mark Z. Jacobson. "Evaluation of Global Wind Power." *Journal of Geophysical Research: Atmosphere* 110 (2005).

Arshadi, Mehrdad, et al. "Energy from Renewable Resources (Bio-energy)." In Christian V. Stevens with Roland Verhé, eds. *Renewable Bioresources: Scope and Modification for Non-food Applications.* Hoboken, NJ: Wiley, 2004: 105–137.

Avery, William H., and Chin Wu. *Renewable Energy from the Ocean: A Guide to OTEC.* New York: Oxford University Press, 1994.

Beijing Declaration on Hydropower and Sustainable Development, Adopted at the United Nations Symposium on Hydropower and Sustainable Development, Beijing, China, 29 October 2004. unhsd.icold-cigb.org.cn/english.html

Bjornsson, Jakob, Ingyar Birgir Frideifsson, Thorkell Helgason, Halldor Jonatansson, Johann Mar Mariusson, Gudmundur Palmason, Valgardur

Stefansson, and Loftur Thorsteinsson. "The Potential of Geothermal Energy and Hydropower in the World Energy Scenario in Year 2020." World Energy Council. 1998. 217.206.197.194:8190/wec-geis /publications/default/tech_papers/17th_congress/3_1_10.asp?mode =print&x=61&y=8

British Petroleum. "Statistical Review of World Energy 2006." 2007. www.bp.com/statisticalreview

Criswell, D. R. "Solar-electric Power via the Moon." *Power Technology International* (Spring 1997): 24–26.

Dunn, S. *Hydrogen Futures: Toward a Sustainable Energy System.* Washington, DC: Worldwatch Institute, 2001.

Energy and Geoscience Institute at the University of Utah. "Geothermal Energy: Clean Sustainable Energy for the Benefit of Humanity and the Environment." 2001. www.geothermal.org/GeoEnergy.pdf

Energy Information Administration (EIA). *Monthly Energy Review.* DOE /EIA Report-0035 (2007/01). Washington, DC: Katherine E. Seiferlein, 2007. tonto.eia.doe.gov/FTPROOT/multifuel/mer/00350701.pdf

European Commission, *Photovoltaics in 2010.* Brussels: AGORES. 2001. www.agores.org/Publications/PV2010.htm

European Commission. *European Union: Energy and Transport Figures.* Brussels: 2005. ec.europa.eu/dgs/energy_transport/figures/pocket-book/doc/2006/2006_energy_en.pdf

Green, M. A., K. Emery, D. L. King, S. Igari, and W. Warta. "Solar Cell Efficiency Tables (Version 24)." *Progress in Photovoltaics: Research and Applications.* 12 (2004): 365–372.

Hammerschlag, Roel. "Ethanol's Energy Return on Investment: A Survey of Literature from 1990–present." *Environmental Science and Technology* 40, 6 (2006): 1744–1750.

Hassan, Ahmad Y., and Donald Routledge Hill. *Islamic Technology: An Illustrated History.* Cambridge, UK: Cambridge University Press, 1986.

Herbert, G., M. Joselin, S. Iniyan, E. Sreevalsan, and S. Rajapandian. "A Review of Wind Energy Technologies." *Renewable and Sustainable Energy Reviews* 11 6 (2007): 1117–1145.

Hoffman, P. *Tomorrow's Energy: Hydrogen, Fuel Cells, and the Prospects for a Cleaner Planet.* Cambridge, MA: MIT Press, 2001.

Huber, Peter W., and Mark P. Mills. *The Bottomless Well: The Twilight of Fuel, the Virtue of Waste and Why We Will Never Run Out of Energy.* New York: Basic Books, 2005.

International Energy Agency. "World Energy Outlook 2006." 2006. www .worldenergyoutlook.org

Komor, Paul. *Renewable Energy Policy*. Lincoln, NE: iUniverse, 2004.

McInnes, Colin R. *Solar Sailing: Technology, Dynamics and Mission Applications*. New York: Springer-Verlag, 1999.

Melosi, Martin V. *Coping with Abundance: Energy and Environment in Industrial America*. New York: Alfred A. Knopf, 1985.

National Renewable Energy Laboratory. *Concentrating Solar Power: Energy from Mirrors*. Washington, DC: Department of Energy, 2001.

Pimentel, D., A. Pleasant, J. Barron, J. Gaudioso, N. Pollock, E. Chae, Y. Kim, A. Lassiter, C. Schiavoni, A. Jackson, M. Lee, and A. Eaton. "U.S. Energy Conservation and Efficiency: Benefits and Costs." *Environment, Development and Sustainability* 6 3 (2004): 279–305.

PowerLight. "Press Release—World's Largest Solar Photovoltaic Power Plant to Be Built with GE Investment and PowerLight Technology." April 27, 2006. www.renewableenergyaccess.com/rea/partner/story?id=44745

Seager, Ashley. "How Mirrors Can Light up the World." *The Guardian* (November 27, 2006). www.guardian.co.uk/business/2006/nov/27/renewableenergy.environment

Shapouri, Hosein, Michael Wang, and James A. Duffield. "Net Energy Balancing and Fuel-Cycle Analysis." In *Renewables-Based Technology: Sustainability Assessment*. Wiley Series in Renewable Resource, edited by Jo Dewulf and Herman Van Langenhove. Hoboken, NJ: Wiley, 2006: 73–85.

Smil, Vaclav. *Energy at the Crossroads: Global Perspectives and Uncertainties*. Cambridge, MA: MIT Press, 2003.

Sørenson, Bent. *Renewable Energy: Its Physics, Engineering, Use, Environmental Impacts, Economy and Planning Aspects*. Burlington, MA: Elsevier Academic Press, 2004.

Tester, Jeffrey W., Elizabeth M. Drake, Michael J. Driscoll, Michael W. Golay, and William A. Peters. *Sustainable Energy: Choosing among Options*. Cambridge, MA: MIT Press, 2005.

Tokimatsu, K., J. Fujino, Y. Asaoka, Y. Ogawa, K. Okano, T. Yoshida, R. Hiwatari, S. Konishi, S. Nishio, K. Yamaji, and Y. Kaya. *Studies of Nuclear Fusion Energy Potential Based on a Long-term World Energy and Environment Model*. 2003. www.iaea.org/programmes/ripc/physics/fec2000/pdf /sep_03.pdf

World Bank. *Tapping East Asia and Pacific's Hydropower Potential*. 2007. web.worldbank.org/WBSITE/EXTERNAL/COUNTRIES/EASTASIA-

PACIFICEXT/EXTEAPREGTOPENERGY/0,,contentMDK:20491251
~menuPK:574044~pagePK:34004173~piPK:34003707~theSitePK:574015,
00.html

World Commission on Dams. "Dams and Development: A New Framework for Decisionmaking." *The Report of the World Commission on Dams.* November 2000. www.dams.org/report/

World Energy Council. "Solar Energy." *Survey of Energy Resources.* 2001. www.worldenergy.org/documents/ser_sept2001.pdf

2

Problems, Controversies, and Solutions

Introduction

In Chapter 1, the renewable and alternative energy technologies in use and under development throughout the world were explored. Although it was acknowledged that a debate over the transition to renewable and alternative energy sources goes on around the world, specifics were not provided. This chapter provides an in-depth examination of the conflicts surrounding renewable and alternative energy, looking at the problems, controversies, and solutions involving fossil fuel versus renewable and alternative energy use. The section on problems addresses issues that have prompted a debate over an energy transition, most of which center on the negative effects associated with continued fossil fuel use. The section on controversies presents the major disagreements of the energy transition debate. The section on solutions focuses on the technical and political tools that could be used to address the problems and controversies. Particular attention will be devoted to U.S. renewable and alternative energy policies. A section on regional and global efforts to provide solutions concludes the chapter.

Problems

As the leading consumer of fossil fuels in the world, the United States has placed a transition to a more reliable source of energy at the heart of energy issues in the first decade of the 21st century. In the face of the growing problems with fossil fuel usage, advocates argue that the most feasible avenue for sustainably meeting the massive and growing energy demand of the world's most powerful economy is to shift to a renewable and alternative energy economy. Of great concern is the finite quantity of fossil fuels, the negative environmental effects associated with fossil fuel use, and U.S. dependence on foreign oil from regions of great political instability.

The Energy Debate

Many factors have prompted the debate over which energy source should be used. Some experts argue that there is a dire need to shift from fossil fuel use to renewable and alternative energy use. This section examines the scientific, environmental, political, and economic problems that have led to this push for an energy transition.

Scientific

Most energy experts agree that, at some point, the world will exhaust its finite supply of fossil fuels. A more contentious issue is when these resources will actually run out. Experts have little disagreement when it comes to the amount of coal available for future use; according to the United States Energy Information Administration (EIA), reserves in the United States alone can provide for current levels of energy demand for up to 250 years (Energy Information Administration 2007a). The remaining reserves of oil and natural gas are a far more controversial issue. Forecasters differ significantly in their predictions of the peak of oil production, known as the topping point. The topping point of oil will happen when the world has consumed exactly one-half of all oil available. This means that the peak of oil production has been reached and that supplies will begin to decline because there are no other oil resources to draw from. The production of natural gas, which is reliant on oil reserves, is closely linked to this production peak. We address arguments for a late topping point in the controversies section of this chapter, but first we turn to the arguments for an early topping point.

Those predicting an early topping point argue that it has either already happened or will happen by the year 2010. Many of these experts are geologists who worked in the exploration divisions of major oil companies and who argue that the bulk of oil that can be derived using conventional methods has been found. A major part of this argument is that the laundry list of geological conditions necessary for oil to be discovered is virtually impossible to meet. All prospective oil reserve locations on land that meet these conditions have been explored, and so it is highly unlikely that very much oil will be discovered on land in the future. Furthermore, it is argued that half of the oil known to be in current reserves will be unrecoverable even through enhanced methods (Deffeyes 2005, 28). Although nonconventional oil deposits, like tar shales and deep water oil, may exist, it is argued that retrieving oil from these locations would require as much or more energy than could be produced from the oil extracted. Such endeavors would be counterproductive because more energy would be used trying to find the oil than could be used from the oil.

Another major aspect of the early topping point argument is Hubbert's curve. M. King Hubbert was a geophysicist who worked for Shell Oil Company. In 1956, by combining the principles of geology, physics, and mathematics, he was able to successfully predict that the United States would reach its peak of oil production in the 1970s. When oil production in the United States did, in fact, begin declining as Hubbert predicted, his prediction for the peak of world oil production in the year 2000 began to gather attention. Although it is unconfirmed that world oil production has peaked, it is likely that we will not know when it has peaked for some years due to the unreliability of data about current oil reserves (Laherrere 2006). The data is unreliable because it is politicized. Oil reserves data is compiled and released by corporations and countries, like those in the Middle East, that benefit from tweaking the numbers.

Environmental

In 1983, the United Nations developed a resolution to create the World Commission on Environment and Development. This commission was later named the Brundtland Commission after its chair, Gro Harlem Brundtland. Its purpose was to develop a report to determine long-term environmental strategies for sustainable development. The report, released in 1987 by the Brundtland Commission, brought to the world's attention the preeminent need for the adoption of sustainable development practices. The principles of sustainability promoted by the commission encouraged the conservation of resources and the reorientation of technology. By contributing to the awareness of the problems with nonsustainable energy sources like fossil fuels, the Brundtland Commission helped stimulate and publicize the debate over the finite quantity of fossil fuels and the other reasons for the need to develop renewable and alternative energy technologies. Energy concerns have been at the forefront of political debates ever since.

To many of those advocating a speedy adoption of renewable and alternative energy technologies, the predicted dates for the topping points of oil, gas, and even coal are largely irrelevant. Environmental concerns and the possibility of tipping points, which can be likened to the domino effect, mean that continued dependence on fossil fuels could cause irreversible damage to the natural world. The consequences of irreversible damage could be devastating. The worst-case scenario is a mass extinction of species, including humans. Before laying out the more recent worst-case scenarios proposed by experts engaged in this debate, it is helpful to look at other environmental and health problems that have been associated with the use of fossil fuel for decades.

Because fossil fuels must be combusted or burned to generate energy, the problem of air pollution has been a concern since the beginning of the Industrial Revolution. Air pollution has mainly been the result of using fossil fuels to generate electricity and to power internal combustion engines, especially those used in the transportation sector like automobiles, trains, and, increasingly, airplanes. Air pollution, in this sense, includes chemicals released from burning fossil fuels. These chemicals are nitrogen oxides, sulfur dioxides, volatile organic compounds, ozone, particulate matter, mercury, carbon monoxide, and lead. Air pollution not only causes thousands of premature deaths per year in industrialized nations, but it also has devastating effects on the environment through the effects of acid deposition, commonly known as acid rain. Acid rain is caused by emissions of acidic compounds, like sulfur dioxide and nitrous oxide, that pollute the environment through precipitation like rain and snow.

Acid rain causes severe damage to lake and river ecosystems, trees, and plants, and it can even cause degradation to human-made buildings. It has also been blamed for reductions in timber production in industrialized nations, leading to economic losses. A growing concern is that the increasing acidification of soil will result in an imbalance of chemicals and nutrients in the soil. Such an imbalance could cause the soil to become infertile. Because plants rely on the nutrients in fertile soil to grow, acid rain can ultimately make it very difficult or impossible for plants to grow. Furthermore, ecosystems can store these acidic compounds in the soil. When rains or floods transport the acidic chemicals to other parts of the ecosystems, there can be negative effects on the environment long after the initial pollution. This delayed effect makes it hard to determine the full impact that acid rain has on the environment, causing us to underestimate the damage. By not being able to gauge the actual effect of acid rain, we risk polluting the natural world beyond repair. Soils and ecosystems can absorb only so much acid rain; if a tipping point is reached, the environmental damage could become irreversible. The death of forests, lakes, and croplands due to acid rain would undoubtedly and clearly have shocking effects on the economy.

Water sources can be polluted by fossil fuel use not only through acid rain, but also through pollution to surface water caused directly by the fossil fuel. An example of this problem is the *Exxon Valdez* oil spill. In 1989, the Exxon oil tanker *Valdez* polluted hundreds of miles of the Alaskan coast causing severe ecological havoc after the ship's hull was damaged from running into shallow water. Pollution of groundwater and drinking water can also be caused, at least in part, by energy exploration practices, including mining and drilling for oil.

Another concern among renewable and alternative energy supporters is the effect of energy exploration on land. Surface mining methods like strip mining, open-pit mining, and mountaintop mining cause permanent alteration of the landscape because the land above the deposits is simply pushed aside by bulldozers and other earth-moving machines. This can severely damage ecosystems and the organisms that live in them, and it requires considerable amounts of water for restoration if restoration is even attempted. Oil exploration and drilling can also disrupt ecosystems and displace the use of arable land. The transport of fossil fuels, once they are discovered, can have detrimental effects on the environment as well. Pipelines can be hundreds of miles long, disrupting the areas through which they run. Additionally, there is the potential for gas leaks and oil spills across the large area of land that pipelines cover. In 2006, an oil spill from a pipeline in Alaska was the result of a pinprick hole that slowly widened as the pressure from the oil pushed against it. It went undetected for several days because the leak formed beneath the snow cover, resulting in the pollution of the environment by thousands of gallons of oil.

There is little debate that air pollution, acid rain, and energy exploration have severe consequences for environmental and human health. Governments all over the globe have enacted numerous policies to address these issues. Although these issues contribute to the push by many scientists and experts for a transition to a more sustainable form of energy, they are not at the forefront of the debate. The major argument of rapid transition proponents is the looming uncertainty of the intensity of the effects of climate change or global warming. Much of the scientific evidence concerning global warming suggests that the process is being accelerated by anthropogenic, or human-caused, greenhouse gas (ghg) emissions (International Panel on Climate Change 2007, 2). These emissions are due almost entirely to the burning of fossil fuels for energy use. The fragile balance of the Earth's atmospheric makeup gives cause for concern about altering the chemicals in the shield that is vital for all life. Increases in carbon dioxide and methane emissions have the effect of trapping heat from the sun like a greenhouse and creating, over time, a rise in temperature that will have dramatic effects on ecosystems. The ability of life on Earth to survive after a certain degree of temperature increase is hardly a point of contention among scientists.

After the possibility of global warming was discovered, the United Nations Environmental Programme (UNEP) and the World Meteorological Organization (WMO) established the International Panel on Climate Change (IPCC) in 1988 to study and assess the potential impact of human-induced warming. Their findings indicated that ghg emissions from fossil fuel usage are most likely the primary cause of the excessive warming that is occurring and that is expected to occur in the future. Wide acceptance of the IPCC's findings is reflected in global agreements like the Kyoto Protocol, signed in 1997 by 166 countries and other governmental entities, which sets binding limits on ghg emissions. Although major greenhouse gas emitters like the United States have not signed the agreement, the general acceptance of climate change is pervasive throughout American society where numerous states and corporations are making their own commitments to reductions in ghg emissions despite the lack of action by the federal government.

The real question is not whether global warming is happening; the general consensus is that it is. The controversy is over how fast it is occurring, how much of a change in temperature there will be, what the real effects of this change will be, and what the impact of nonhuman-related causes may be. Because of this uncertainty, global warming has also become a contentious issue: On one side are those who would advocate a rapid transition away from fossil fuels, prompted by government intervention through policy mandates; on the other side are those who would opt for a gradual transition through market mechanisms or no transition at all. We turn first to the arguments for a rapid transition.

Several factors have warranted concern among scientists, policy makers, and citizens alike. With the continued combustion of fossil fuels and hence the emission of greenhouse gases, several scenarios for the effects of warming have been explored through scientific modeling. The degree of warming that will occur could be too high for many ecosystems to survive because the rate of the planet's current and future warming could be too fast for ecosystems to adapt to it. The result could be the die-off of entire ecosystems like tropical rain forests and coral reefs. Experts are already predicting that, with current levels of warming, coral reefs are in danger of severe degradation and possible extinction (Buddemeier, Kleypas, and Aronson 2004). This loss of biodiversity would have many dire effects on human and natural populations. Just from an economic standpoint, the loss of resources like the rain forests would be devastating for the regions that depend on them. From a global perspective, a significant loss of ecosystems could have even more serious effects on humanity. Furthermore, specific ecosystems, like the rain forest, serve to provide global benefits that are absolutely necessary for the survival of humanity. For example, we rely on rain forests and ocean ecosystems for oxygen to breathe. On the flip side, warming would not only destroy ecosystems but also could increase the geographic range of survival for certain organisms, like mosquitoes. The result would be the spread of diseases like malaria to parts of the world where they were not previously a problem.

Rising sea levels are also a concern. As global temperatures increase, glaciers and ice caps are expected to melt, causing a gradual increase in sea levels over time. Recent studies have shown that ice caps and glaciers are indeed melting. In fact, the rate of melting is much higher than was originally thought by experts (Stroeve et al. 2004). Sea levels, which originally were not predicted to rise significantly for another millennium, are now predicted to rise anywhere from several inches to several feet in less than one hundred years (International Panel on Climate Change 2007a, 13). Much of the world's infrastructure and many major cities are located on the coasts. The effect of rising sea levels on cities like New York; Tokyo; Shanghai, China; Rio de Janeiro, Brazil; and Los Angeles would be particularly disastrous in terms of human health and economic expenses. A new classification of environmental refugees would be seeking protection inland, causing strain to already stressed ecosystems and cities.

Rising sea levels, coupled with predictions for more frequent and intense storm systems, would not only put immense strain on the insurance industry but could bankrupt it altogether (Leggett 2005, 72). This in turn could cause the collapse of global financial markets because production would be halted and capital to repair damages would be nonexistent. The same problems that would plague the insurance companies are also a threat to the agricultural sector. Flooding and droughts, the effects of heat on crops, and the proliferation of pests could dramatically reduce global food supplies (International Panel on Climate Change 2007b, 6). Similarly, droughts would result in the compounding of an already stressed water supply. Effects like this are being felt in rural areas in developing countries where disadvantaged populations must rely on international assistance to cope with such problems (Revkin 2007, 1). Any one of these effects, and certainly all of them combined, would add to the possibility of water-related conflicts and societal instability. The potential for violence is beyond what experts have attempted to describe when presenting the global warming scenario.

Perhaps the most worrisome scenario is the possibility that these effects will occur much sooner than predicted and that the warming could become irreversible. The chances of naturally occurring amplifying feedbacks create the potential for an apocalyptic situation like something dreamed up for a Hollywood movie. Amplifying feedbacks refer to the release of naturally occurring greenhouse gases, many of which are currently dormant due to low temperatures. A rise in temperatures could result in the stimulation of emissions from sources like methane hydrates found in the oceans and greenhouse gases locked up in permafrost. The so-called methane burp hypothesis, which refers to the sudden and immense release of methane from melting hydrates in the ocean, contends that past mass extinctions may have already resulted from accelerated warming caused by methane, which is a greenhouse gas that is many times more powerful than carbon dioxide. If these methane hydrates were destabilized again, they could cause catastrophic warming as well as a dramatic change in the makeup of atmospheric gases, affecting the ability of life dependent on oxygen and carbon dioxide to survive. Additionally, experts predict that warming could cause the acceleration of decomposition in soil, causing it to switch from a carbon sink to a carbon source. The loss of other carbon sinks, like the Amazon rain forest, would affect the Earth's ability to naturally balance atmospheric levels of oxygen and carbon dioxide. Coupled with anthropogenic additions of greenhouse gases (like carbon dioxide) to the atmosphere, such an occurrence would result in a situation that models have not been able to fully realize. These amplifying feedbacks could result in a rate of warming much faster than predicted, with much more intense effects. If the impacts of global warming reach the worst-case scenario, the trend will become irreversible, and humanity and civilization as we know it will cease to exist.

While there is much uncertainty about the actual results that warming will have and when, several experts agree that the precautionary principle should be employed. The precautionary principle states that, if the impacts of an action or inaction are extreme and irreversible, even if there is inconclusive scientific evidence about the risk for disaster, measures must be taken to prevent the impacts from occurring (United Nations 1992, 10). The facts that fossil fuels are finite and that the world population is growing and developing mean that a new energy source has to be developed at some point. The risks from global warming make the case for developing that energy source as soon as possible. Renewable and alternative energies have the potential to alleviate the natural world from most of the degradation it undergoes now due to the dominance of fossil fuels in the energy market.

Political and Economic

Other issues that have contributed to the debate over the need for renewable and alternative energy sources involve political and social problems associated with fossil fuel use. The natural geographic location of raw fossil fuel resources throughout the world, particularly oil, is very centralized and has resulted in many negative geopolitical consequences. An example of this is the latest war between the United States and Iraq, which has been attributed to the need for a stable oil supply, albeit under the guise of national and global security from weapons of mass destruction (Piven 2004, 14). Incidents like the disruption of natural gas supply in 2006 from Russia to the recently annexed province of Georgia have caused tension in an already volatile region (Champion 2006, 1). Political scientists contend that tension between the United States and Venezuela would not be as severe were it not for the large reserves of oil Venezuela holds (U.S. Government Accountability Office 2006, 2).

As the globalization of world economies and markets speeds up the development of highly populated countries like China and India, these nations begin to compete with already developed countries for their share of fossil fuels. Because fossil fuel supplies are limited, competition, especially as these resources become scarcer, could escalate into conflicts where the use of military force is entirely possible. Furthermore, either countries without significant fossil fuel resources could become too dependent on those that have fossil fuels, making their economies much more vulnerable to price fluctuations, or they could become aggressive in their efforts to secure energy for their people (Center for American Progress 2006, 9). Most developing countries are just entering the fossil fuel age, meaning that the worst effects of mass consumption are yet to come. To prevent the global consequences of fossil fuel use, there must be a transition to a reliance on renewable or alternative energy by all countries, industrialized and developing. Industrialized nations have caused many problems already through their fossil fuel combustion, and developing countries cannot rely on fossil fuels without the same negative local, regional, and global implications.

To address the global effects of fossil fuel use, efforts at the worldwide level must be made because many of the effects of global warming have an impact on what is called a pure public good. A pure public good is a resource that can be used by everyone at the same time and that no one can be stopped from using, like the atmosphere. Consequently, the misuse of this good, such as the pollution of it, affects the quality of the good for all users to some degree. For instance, if the United States and China burn fossil fuels and emit greenhouse gases into the atmosphere, after the levels of those gases reach a certain point, the entire world, not just the United States and China, have to bear the burden of the effects of the increased warming. The United States, however, has no incentive to bear the costs of reducing its ghg emissions, particularly if another country, like China, is going to continue to emit them. This phenomenon is called the tragedy of the commons. The only way the tragedy of the commons can be avoided in this situation is if both countries agree to reduce their emissions enough to eliminate their contributions to global warming. In the case of renewable and alternative energy, efforts at a transition will be more successful if all of the most energy-intensive countries in the world make an agreement to switch. As such, the problem of getting the major energy consumers of the world to sign a binding agreement could hinder the transition to renewable energy.

Controversies

This section examines the other side of the energy debate, which does not support a shift away from fossil fuels. The critics of a transition to a renewable and alternative energy economy focus on several important issues. Although proponents of renewable and alternative energy have answered many of these concerns, they are important to discuss nonetheless. By looking at the situation from as many angles as possible, we can develop the most feasible solution to the problems. First, we address the problems associated with the renewable and alternative energy technologies described in Chapter 1. Then we present the alternative side to the debate over the problems with fossil fuels just presented.

Controversy: Viability of Renewable and Alternative Energy Technologies

While using renewable and alternative energy has many advantages, the technologies also have some drawbacks and hurdles to overcome. Although each technology has specific problems to surmount, many of the technologies share similar obstacles to use on a mass scale. If we are to rely on renewable and alternative energy sources to meet our energy needs, the technologies must be viable for widespread use. In this section, we discuss the individual and collective barriers to the use of these technologies on a mass scale, along with the possibilities for overcoming them. These barriers include problems with the storage and distribution of energy produced from renewable sources, the costs of the technology, technicalities, environmental effects, and social and political barriers. A separate section presents the particular problems associated with nuclear power.

Storage and Distribution of Energy

With the exception of biomass and nuclear power, a recurring problem with renewable and alternative energy is the need for the storage and distribution of energy, although the latter is typical for most energy sources, including fossil fuels. While the decentralization of renewable energy sources can alleviate problems like those associated with its geopolitical characteristics and equity among less developed countries, it can pose a problem for areas that require excessive amounts of energy, like major metropolitan areas. This is because the quality of energy sources can vary by region. For example, although solar is inexhaustible, it is subject to seasonal fluctuations of light and has much wider availability in some regions than in others. The Southwest region of the United States has considerable solar resources, and the Great Plains region has excellent wind potential, but the energy from these areas would have to be transmitted across the country if renewable energy is to provide for the needs of areas like New York City or Atlanta. For renewable energy from these regions to meet the energy needs in distant cities, the energy must be transmitted as electricity by wire or stored in another medium as potential energy and transported for distribution. If stored in a medium, some kind of technology has to be used to utilize the energy from that particular medium. For example, if renewable energy were used to make hydrogen fuel that was transported by tanker to New York City, hydrogen fuel cells would be required to use the potential energy stored in the fuel.

Even in areas having an abundance of renewable energy, natural fluctuations, like those due to weather and seasons, create the need for storage. For example, while there is more than enough solar energy in the Southwest to provide power for the area year-round, the sun does not shine at night, and monsoons in the summer prevent solar radiation from being able to provide energy directly 24 hours a day. Excess energy must be stored for use at night and during other times when there is little sunlight. These variations in supply are common for all renewable energies except biomass. Wind is a much more variable and localized energy source than solar. The quality of wind can differ greatly from region to region, which causes a need for energy storage if it is to be used on a large scale. Geothermal is often not located near the areas that need the most power. Ocean energy is also very localized, requiring the transmission or transport of electricity over large distances. Technologies like ocean thermal energy conversion (OTEC) may have to operate far offshore, which complicates the transport of energy through transmission lines. Like other renewable energies, ocean energy would benefit from a viable storage technology. Hydroelectric dams are even less modular than most renewable energies; they must be built in very specific places. Yet large-scale hydroelectric dams have enjoyed success as energy providers. With financial investments in such projects, transmission lines have been built to transport hydroelectricity to city centers. This fact alone makes a case for the possibilities of overcoming issues with the distribution of energy produced from renewable sources.

As discussed in Chapter 1, hydrogen has great potential to act as a storage medium for renewable energy. By converting renewable energy into hydrogen and storing it in a fuel cell, it can be transported as a highly efficient energy source and used in a variety of energy applications. It makes full exploitation of the most concentrated renewable energy power densities possible. This highly modular characteristic could eliminate the need for distribution via cumbersome transmission lines for large grid applications. Hydrogen is efficient and reliable. It is clean and emits nothing but water when combined with oxygen to create electricity in fuel cells. Problems with hydrogen fuel technologies do exist, however. It is not yet economically viable or at least cost competitive with fossil fuels. There is a lack of infrastructure in place for the implementation of hydrogen fuel technologies, especially in the transportation sector. There are also safety concerns about its flammability. While it has a lower flammability limit than gasoline or oil, it has a much greater range of flammability. Technology to effectively store hydrogen fuel is still being developed. As mentioned, however, impediments to the use of hydrogen are more financial than technical. The main barrier to expanded energy production from either of these sources is the perceived monetary costs. Hydrogen production requires significant amounts of energy, and materials for fuel cells are expensive enough to prohibit the expansion of this technology. These costs may be more perceived than real, though, as will now be discussed.

Costs

The costs of renewable and alternative energy are probably the most prohibitive factors to their development on a wide scale.

Although renewable and alternative energies can actually pay for themselves in the long run, they have extremely high initial capital costs in the short term. As a result, they are not as cheap at first for consumers to buy as fossil fuels, making the demand for renewable and alternative energy lower than demand for fossil fuels. This, in turn, discourages investment into renewable and alternative energy by the private sector, which is essential if the technologies are to be produced on a large scale. Cost has proved inhibitive for the development of solar energy in particular (Perlin 2002).

A couple of contributing factors have ensured that fossil fuels are the most cost-competitive energy source on the market. First, subsidies and tax breaks from federal governments have kept the price of fossil fuels down, despite the high costs of exploration and extraction (Myers and Kent 2001, 65). While renewable and alternative energies enjoy some financial assistance from the government, they do not receive as much as fossil fuels, nor have they received it for as long. This gives fossil fuels an unfair economic advantage over more sustainable energy sources.

Second, arguments have surfaced that many of the costs of fossil fuels and the benefits of renewable energy and energy from hydrogen are not accounted for. These costs and benefits are known as externalities. "Externality" is a term used by economists to describe a cost or benefit that is not calculated into the costs or profits of production. Often such costs and benefits are paid for or enjoyed by the public and by the government, rather than the industry. Negative externalities are associated with costs, and positive externalities are associated with benefits. For example, if photovoltaics (PVs) are used instead of a coal-fired power plant to generate electricity for a city, the costs and profits are calculated based on the energy production. In that respect, power produced from photovoltaics is much more expensive than power from the coal-fired power plant. This is due to the amount of technology needed to produce the same amount of energy as the coal plant and the high cost of that technology. The result is that the capital costs of using solar energy are much higher than the cost of the coal plant's construction and the coal itself. If the decision to choose solar or coal as an energy source is based on the costs of the two projects, coal would be chosen because it is much more inexpensive and turns a quicker and more significant profit than solar. Coal, however, contributes to air pollution and acid rain, whereas solar does not. If the costs borne by society for negative externalities associated with the air pollution and acid rain caused by coal-fired power plants were calculated into the cost of production, then the costs of the coal-fired plant would be much closer to, or perhaps greater than, the price of solar power. If positive externalities, like cleaner air, improved environment, and energy security, were factored into the profits of solar energy production, then it would become even more cost competitive with coal and other fossil fuels. Unfortunately, calculating these costs and benefits is very difficult because it is hard to put a price on clean air and healthy ecosystems. As a result, most economic analyses simply omit these externalities from cost-benefit ratio calculations, causing a distorted view of the real costs of using renewable energy over energy derived from fossil fuels. But as energy prices rise, particularly oil, renewable and alternative energies become more cost competitive even when externalities are not accounted for (Odell 1999, 37). This glitch in our economic system can be overcome through policies designed with these facts in mind. A detailed discussion of this is provided in the section on solutions.

Technicalities

Like most technologies, renewable energy technology is in need of further development and refinement. The conversion of energy from renewable sources for more practical applications is still not as efficient as most critics argue it should be to be viable for large-scale use. Most solar technologies have much potential for improvement. Although wind energy is fairly efficient, transmitting the energy over long distances to urban centers can result in efficiency loss. Even biomass, which does not have the energy storage problems of other renewable energy sources, is not as efficient an energy source as fossil fuels (Hammerschlag 2006, 1746). This is because the density of energy intensity found in fossil fuels is much higher than it is in biomass. The result in each of these cases is that larger quantities of the original energy sources would have to be used to achieve the same level of energy production as fossil fuels. This creates a greater problem for biomass than for inexhaustible energy sources like wind and solar. Due to inefficiencies in the conversion process, ocean energy may be unable to provide enough energy to make it suitable for largely populated areas. However, the more that these technologies are researched and developed, the quicker the improvements needed for efficiency will be found.

Other technical issues involve the vulnerabilities of the technology to the elements. Similar to offshore oil drilling, offshore wind and ocean technologies are vulnerable to damage in storms, which can lead to interruptions in energy supply. Large- and small-scale hydropower projects are at risk from river flooding during excessive rainfall or snowmelt. And dams have to contend with the issue of silt buildup. In some cases it may take hundreds of years for the built-up silt to render the dam unusable for the production of hydropower. Depending on the flow of the river and the amount of sediment it carries, however, silt buildup can occur much more quickly.

Environmental Effects

While the environmental effects of renewable energy are miniscule compared to fossil fuels, they do have an impact on the environment, some much more than others. Being aware of these impacts makes it easier to minimize them and to make more informed decisions about energy development. Until these technologies are developed further, complex technologies, like photovoltaics made from silicon products, will continue to require a large amount of manufacturing energy. If this energy does not come from a sustainable source, the one-time production process of these technologies could contribute to pollution and emissions, though much less when compared to the life of fossil fuel technologies. Although biomass can create carbon sinks that offset its emissions, it still releases carbon dioxide when combusted. And it has been found that reservoirs created by large dams emit small amounts of greenhouse gases due to the accelerated breakdown of ecomatter.

The use of land and disturbances to ecosystems and wildlife are major concerns. Wind farms, central solar power (CSP) systems, and especially biomass and large-scale hydroelectric projects can require a considerable amount of land. In fact, this is considered one of the biggest drawbacks of biomass. Competition for land between energy sources and the agricultural sector could potentially offset agricultural development, particularly in the case of biomass. If we rely too much on biomass to provide for our energy needs, interruptions in the food supply are also a possibility. Additionally, to expand the availability of land for the cultivation of biomass crops, deforestation could occur, which could then cause the loss of those carbon sinks creating a negative trade-off and potentially increasing carbon dioxide in the atmosphere. The potential for this deforestation to cause a host of other ills—like soil erosion, desertification, loss of biodiversity, and the siltation of rivers—would be greatly amplified. Large-scale hydro has been known to cause significant disturbances in land and river ecosystems through the flooding that occurs with the creation of the reservoir. Dams cause changes in temperature and in the chemical content of water, and they interfere with the quality or flow of water downriver, which can also have detrimental effects on river ecosystems. Ocean energy has the potential to disrupt ecosystems and wildlife, including interference in the transport of sediment by waves. Offshore wind and ocean technology requires the use of areas in the sea, which could interrupt nautical navigation for humans as well as fish and bird populations. Both onshore and offshore wind projects have been found to be hazardous to bird populations, though technological adjustments are expected to mitigate this effect (Chambers 2000, 14). Dams also block the migration of fish populations like salmon, making it difficult or impossible for them to return to spawning areas and often pushing them to the point of extinction. The environmental concerns related to geothermal energy center on the extraction of the resources. Although the risks are low, extraction has the potential to cause seismic reactions from drilling and land subsidence from the extraction of water if replacement fluids are not injected into the reservoir.

Biomass also has the potential to deplete soil nutrients necessary for agriculture, though, if used as a rotation crop in agricultural operations, it could have the opposite effect. Similar to the effect of agriculture on the environment, biomass crops could cause fertilizer contamination of water, negative impacts from monoculture, and other pollution related to the growing and harvesting of crops. The use of water is a concern in the cases of biomass and geothermal. The cultivation of biomass crops uses much more water than other renewable energies. Although it is a minimal concern, certain types of geothermal exploration and extraction are also water intensive.

Social and Political Barriers

Generally, support for renewable energy and energy from hydrogen is high. However, these sources still have social and political issues that must be dealt with. There has already been dissent over the locations chosen for some types of renewable energy developments. The dissent stems from what has come to be known as the NIMBY phenomenon, NIMBY being an acronym for "not in my backyard" and referring to the opposition of area residents to a development in their geographical location but not in someone else's. A perfect example of NIMBYism is the protest by residents of Cape Cod over the Cape Wind Project. Off the coast of Cape Cod, Massachusetts, the Cape Wind Association is developing the first offshore wind farm in the United States, called the Cape Wind Project. High-profile property owners in the area, including Robert F. Kennedy Jr., are hotly contesting the project because of a perceived negative effect on the seascape, the local economy, and the ocean, which is considered a global commons (Kennedy 2005). This lack of local acceptance for the project could cause delays in its construction or even halt it altogether. Although this example highlights a rejection due mainly to aesthetics, wind turbines can also be noisy, making them less socially accepted in urban areas. This creates a difficult hurdle for renewable energy in particular because of its localized nature. Certain sites are naturally better for renewable energy development than others because of the high concentration of renewable energy found there.

While the NIMBY phenomenon seems to be one of the biggest social and political problems for renewable energy development, other problems are related to specific technologies. The conditions for working on turbines in wind farms are hazardous to workers, although this is also true of many fossil fuel extraction and utilization technologies. The use of biomass or biofuels, as illustrated in the preceding section, could cause the energy needs of society to come into direct competition with food and timber production. The creation of reservoirs can cause a severe displacement of human populations, as the Three Gorges Dam project did in China on the Yangtze River, which forced 1.2 million people to relocate. With regard to the remaining undeveloped hydropower potential in the world, the majority of potential exists in areas of political instability, making it difficult to get a project online, much less one that is safe from interference. However, most proponents of renewable and alternative energy do not support the development of large-scale hydro projects because of the myriad issues associated with them.

Controversy: Nuclear Power

Some argue that the problems associated with nuclear power have killed it as a possibility for solving our energy woes. Recent events in developing and even in developed countries, though, have shown that nuclear energy has yet to be abandoned as an energy option. While far fewer issues are associated with nuclear fusion than with nuclear fission, fission is the only viable way to create nuclear energy, and so its disadvantages must be taken into account. This section examines the problems associated with both fission and fusion, paying particular attention to the problems associated with fission. Unlike renewable energy, nuclear energy does not have the problems with storage and distribution; therefore, our discussion begins with costs.

Costs

Like renewable energy, nuclear power has high initial capital costs. Although nuclear power is used commercially, the economic benefits of it are misleading. First, it requires considerable initial capital investment that is often subsidized by the government and hence the taxpayers. In reality, nuclear power is not economically competitive with fossil fuels on its own. The capital investment needed to build a fusion reactor core, once the technology is viable, is even higher than that required for the construction of fission reactor cores because fusion reactor cores are much larger and more complex. Second, the government subsidizes the insurance for the operation of such plants so that they are affordable to run by the private sector. Although the risks are low that there could be a reactor meltdown or radiation leak, the costs to insure against the astronomical harm are too high to make nuclear power an enticing investment for the private market. Again this burden is passed on to the taxpayers. Third, the costs of waste storage and disposal, maintenance, plant decommissioning, and reparations in the case of a meltdown also make nuclear power very expensive. Finally, as discussed below in the next section, if there is a uranium shortage and uranium supplies decrease, demand will continue to increase as will the price, making nuclear energy even less economically viable. There is little debate over the costliness of nuclear energy, but many supporters of nuclear energy view the benefits, especially those for the environment and national security, as outweighing the economic costs.

Technicalities

Both fission and fusion technologies have issues to overcome. Fusion reactions are much more difficult to achieve than fission reactions, as evidenced by the fact that nuclear energy produced by fission became commercially viable decades ago. The fact that energy from fusion reactions has yet to be harnessed and used is discouraging, and the lack of a breakthrough has some countries phasing out fusion research and development funding. Nuclear fission requires the use of uranium. But with an impending shortage that has been estimated to occur as early as 2015, nuclear power will become an undependable source of energy (Jameson 2005). Continued support for nuclear energy is driven by hopes that new technology will be developed that can create nuclear reactions without uranium.

Environmental Effects

Despite the environmental benefits of nuclear power's lack of emissions, the use of nuclear fission, though not so much nuclear fusion, is still a danger to the environment and human health. The Three Mile Island and Chernobyl accidents are proof that things can go wrong in nuclear reactors and that the effects can be potentially devastating. Not only can lives be lost instantly (30 people died immediately after the Chernobyl accident), but the radiation released can lead to more deaths in the future by causing higher rates of cancer. Radiation exposure and poisoning are damaging to ecosystems as well. While the safety records of most nuclear reactors are very good, the risks and consequences of a meltdown or an explosion have made a strong case for the elimination or reduction of the use of nuclear power altogether.

Environmental concerns about radioactive waste have also contributed to the negative perception of nuclear power (Riley 2004, 94). The permanent storage of nuclear waste from current nuclear operations has been a contentious issue for decades. Even after Yucca Mountain in Nevada was designated as the permanent storage site, numerous delays related to litigation and expanded study of the area have prevented it from accepting nuclear waste. Radioactive waste produced from nuclear power generation must be stored for at least 15,000 years without leakages to protect the environment and human health from harmful radiation exposure. It takes thousands of years for the radiation in the waste to subside. Other environmental damages include thermal pollution to nearby waterways by nuclear power plants and disturbances to the ecosystems from uranium mining.

Social and Political Barriers

Although support for minimizing the use of fossil fuels as an energy source is high, support for nuclear power is low and support for renewable energy is mixed. As mentioned in the controversies section, despite the lack of nuclear reactor accidents since Chernobyl, nuclear power is perceived to be dangerous and risky. This fact has contributed to the halt of licensing of any new nuclear reactors in the United States since the 1970s. Though using nuclear waste to make a bomb has not happened yet, there is always that concern. Nuclear waste would be extremely dangerous if the wrong people acquired it. Nuclear power is so expensive and requires such an investment and technical know-how that it can be used only by industrialized and politically stable countries where the risks are accepted by the societies. The NIMBY phenomenon affecting renewable energy sources contributes to the social barriers for nuclear energy as well. Most communities are not supportive of building a nuclear reactor right in their own backyard.

Controversy: Energy Conservation

Chapter 1 presented a discussion of the potential for energy conservation and efficiency to alleviate our dependence on fossil fuels. Although most experts advocate energy efficiency, some argue that energy-efficient technologies would actually raise energy consumption rather than reduce it. As energy efficiency has grown, so has global per-capita energy consumption, which means that increased use is not just due to population growth. The argument is that increased energy efficiency leads to greater energy consumption because it increases the opportunity for even greater use of energy-intensive products. Furthermore, the more efficient technology becomes, the more people can do and the faster they can do it, leading to more energy consumption. Basically, energy efficiency cannot save energy if it causes more people to use energy that they might not have used otherwise (Huber and Mills 2005, 114). For example, suppose a family saves money using energy-efficient appliances and lighting and then uses that money to fly somewhere on vacation. In that case, they are using more energy than they would have used if they were still using inefficient appliances and lighting. There are, however, several flaws to this argument. It assumes that the family would not have flown somewhere on vacation without money from the energy savings, and there is little evidence to support this hypothetical assumption. Furthermore, the argument that more efficient technologies lead to greater use of those technologies, offsetting any energy savings, assumes that the less efficient technologies would not be used by the other people using the technology for the first time. The argument that there is a correlation between an increase in energy efficiency and an increase in per-capita energy use does not prove causation. The correlation could exist simply because of a correlation between the creation of more energy-intensive products and the number of people using those types of products or any similar but less efficient product.

Controversy: Oil Topping Point

As previously mentioned, there is much controversy over the actual date for the oil topping point. Those predicting a late topping point assert that oil production will start to decline no sooner than 2037. Peter W. Huber and Mark P. Mills argue in their book, *The Bottomless Well: The Twilight of Fuel, the Virtue of Waste and Why We Will Never Run Out of Energy,* that the oil topping point is completely inconsequential (Huber and Mills 2005, 181). The natural gas topping point has been predicted to occur as early as 2030. While disagreements abound over the exact date that the finiteness of fossil fuels will be fully realized, virtually all experts concede that it will happen eventually. This latter point, taken alone, points to the need to develop technology that will harness the inexhaustible resources of wind, solar, hydro, geothermal, and biomass before society is affected by the squeeze of limited fossil fuel resources.

Related to the controversy over the oil topping point are some insightful arguments against the utility of energy forecasting that were mentioned in Chapter 1. Dr. Vaclav Smil exposes the historical inaccuracies of energy forecasting and the sensitivity of modeling results to "unanticipated turns of events" in his book, *Energy at the Crossroads: Global Perspectives and Uncertainties* (Smil 2003, 172). He argues that the unmanageable number of variables and their infinite number of interactions with each other are impossible to account for and ensure that forecasts and predictions based on models will continue to be incorrect, often drastically so. An examination of previous energy consumption forecasts seems to make Dr. Smil's point. For example, the two most widely cited forecasts of the late 1970s, those of the World

Energy Conference of 1978 and the International Institute for Applied Systems Analysis, overestimated the use amount of energy consumption in 2000 by no less than 30 percent in their most conservative estimates (Smil 2003, 144).

While Dr. Smil does not deny that energy demand and consumption will continue to increase, he argues that it is futile to utilize all of our resources trying to determine how much they will increase. He contends that the solution to the problem of inaccurate predictions is to shift our efforts toward normative forecasting. Normative forecasting focuses on what should happen rather than what will happen, thereby encouraging the formulation and implementation of proactive solutions that concentrate on creating the results desired (Smil 2003, 178–180).

Controversy: Global Warming

Few, if any, climatologists believe that warming is not occurring. The scientific data to support this belief has been overwhelming. Although the vast majority of climatologists agree that warming is occurring, there is debate over the specifics of the theory and the actions that they warrant ("Global Warning . . ." 2007). Some scientists are concerned that IPCC projections will be inaccurate (Tennekes 2007). Similar to Dr. Smil's argument, they argue that there are too many factors to make an accurate prediction of warming or its effects. Several scientists point to data that contradicts the IPCC assessment of the causes of global warming. Essentially, they argue that global warming is due more to natural causes than to human activities (Khilyuk and Chilingar 2006, 899). Several scientists argue that the causes are not known one way or the other. Dr. Richard S. Lindzen, a professor at the reputable Massachusetts Institute of Technology, is one of the most vocal proponents of this position. Lindzen has contributed multiple times to IPCC studies and publications about global warming. He has remained a vocal critic of the IPCC and any theories (Lindzen 2005). A handful of scientists even believe that warming will have a positive effect on the environment and on humanity (Michaels 2003). As indicated in the section on global warming under problems, many experts and politicians believe that the precautionary principle should be used when making decisions about what actions to take. Despite any doubts about the probability of global warming causing negative consequences, it is worth taking the measures necessary to prevent the possible disasters. By analogy, even though we do not expect to be in a fire, we take precautionary measures to save our lives through the provision of fire escapes and fire stations.

Controversy: Economic and Political Problems

The major economic and political controversies surrounding renewable energy are closely tied to one another. In the United States, fossil fuel companies have considerable political power, and many government officials have vested interests in the success of these companies (Leggett 2005, 125). This results in the passage of more favorable legislation as well as greater funding for fossil fuel development than for renewable and alternative energy development (Myers and Kent 2001, 70). Furthermore, economic considerations of the effects of renewable and alternative energy policy continue to hinder the passage of strong legislation in favor of renewable and alternative energy.

As already discussed, the initial capital costs of investing in renewable energy development are high compared to fossil fuels. Thus, there are two positions on how a transition to renewable and alternative energy sources should occur. The first position holds that the government *should allow market forces to provide the impetus* for the transition. This approach relies on industry and corporations to take the initiative to produce energy from renewable and alternative sources when it becomes more profitable for them than producing energy from fossil fuels. The second position contends that government *should intervene in the market by passing policies that promote or mandate renewable and alternative energy use*, thus speeding a transition to the use of these energy sources more than the market alone could do.

The controversy stems from an age-old political debate over the role of government in free markets. Those who hold the first position argue that economic efficiency has to be sacrificed if governments intervene to encourage the market to act one way or another. Economic efficiency is reached when goods are produced in the market at the lowest possible cost to the producer. The production of a supply of goods in a perfect market is the direct effect of the demand for the good. As discussed, in the case of renewable energy, the demand is not high enough to compel private interests to make the scale of investment needed to drive the costs of renewable energy technology down and to increase demand. Over time, if allowed to function without interference, demand would grow to the point at which production would reach the scale necessary for prices to drop enough to encourage mass demand. The problem with this scenario is that there is no guarantee that the market will work in time to prevent the possible irreversible ecological destruction caused by fossil fuel use or the economic shocks that an oil shortage would cause throughout the world. But if economic efficiency is sacrificed through taxes on fossil fuel use, subsidies for renewable energy, and even fluctuations in the supply of energy, there is a possibility for severe economic consequences that could affect the high quality of life in industrialized countries like the United States. This is the main concern of those who promote a free market solution. The risks associated with an economic downturn are of greater concern to them than the possibility of irreversible environmental degradation. There is a growing consensus, as reflected in the increasing number of renewable energy initiatives taken by governments around the world, that government intervention is necessary at least to some degree. Taking this into account, we turn now to the policy solutions available and to the broader advantages of renewable and alternative energy solutions.

Solutions

Chapter 1 offered a list of the advantages associated with individual renewable and alternative energy sources. In this section, we examine the broader advantages that renewable and alternative energies have to offer the greater society. We then provide a discussion of the United States' renewable and alternative energy policy. Finally, we provide an overview of global and regional efforts to find solutions. Chapter 3 provides an in-depth look at the worldwide perspectives of renewable and alternative energy development.

Broad Advantages of Renewable and Alternative Energy

The broad advantages associated with renewable and alternative energy can be attributed to many characteristics of these sources. We now discuss in detail the positive effects of the decentralized nature of renewable and alternative energy sources. Then we address the economic benefits that the industry can bring to the world in the form of employment opportunities and energy security. Finally, we explain how renewable energy can contribute to gender equity, particularly in developing countries.

The abundance of renewable and alternative energy potential creates a more regional and, hence, more stable supply of energy than fossil fuels, which are imported by most of the world from select regions. Discussed in more detail in Chapter 3, this decentralization of energy resources would not only create more stability, but it could also result in more autonomy of less powerful and less wealthy countries. The potential for renewable and alternative energy to enhance equity in the world energy markets is enormous. One-third of the world's population does not have adequate electricity to meet their needs (IEA 2006, 567). This is partially due to the fact that it is too expensive or difficult to build the infrastructure to transmit electricity over great distances. Renewable, more so than alternative, energy development in nonindustrialized Third World countries and in more rural areas has the potential to result in a more equitable distribution of the benefits of electricity use. Renewable energy developed on a small scale can provide the localized energy needed to provide power to people in less developed areas. This could reduce reliance on fossil fuels in these parts of the countries before it has a chance to take root. Furthermore, the adoption of renewable and alternative energy would allow for the social and environmental benefits of restoring forests and other ecosystems that local people rely on to provide the biofuel that is their primary source of energy, as it is in many developing countries, especially in Africa (Energy Information Administration 1999). The benefits of reducing deforestation are numerous and include the restoration of carbon sinks, which could help to alleviate the effects of global warming.

Economically speaking, the decentralized characteristics of renewable and alternative energy would reduce the costs of transmission as well as the transport of raw fuels. Because renewable sources like solar, wind, and water are virtually free for the taking, the main costs of production would be the initial capital investment in technology and infrastructure. Such operations would essentially pay for themselves once the initial investment is recouped. When examined from a long-term standpoint, renewable energy use could save billions of dollars compared to fossil fuels. These savings do not include the health and environmental costs associated with fossil fuels, which can be hard to measure because it is difficult to fully establish causality in every case. While the exact numbers are difficult to calculate on a global scale, there is little doubt that a considerable amount of money could be saved from a reduction or the elimination of fossil fuel use (Center for American Progress 2006, 12). As already mentioned, many benefits are associated with good health that cannot be assigned a dollar value. This may not apply as much to nuclear power because the cost of one or more nuclear accidents has the potential to far exceed the costs of using fossil fuels to the society and the environment. This fact contributes to the lack of development of nuclear power in many Western nations.

Renewable energy provides a great opportunity for economic growth in terms of new jobs, new industries, and new wealth in both industrialized and developing countries. Studies have shown that in some instances renewable energy provides more economic opportunities than the use of fossil fuel sources do (Center for American Progress 2006, 10). Many renewable energies offer the added benefit of being able to create more jobs per unit of energy generation than fossil fuels (MacGill, Watt, and Passey 2002). These jobs are diverse, ranging from extensions to existing agricultural and forestry activities through specialized engineering and electronic functions. Many of them are located in rural areas where the need for employment is the greatest (Saddler, Disendorf, and Denniss 2004, 77). The economic advantages of renewable energies are particularly applicable to women in the developing world. In nonelectrified rural areas of developing countries, women do most if not all of the housework, in addition to other jobs required to provide for the family (Clancy, Oparaocha, and Roehr 2004, 11). This creates an unequal division of labor between women and men because women work as much as men outside the house and then do the majority of the work required to maintain the household. Because renewable energy has the potential to bring electricity to such rural areas, it also has the potential to lighten the workload of women in the home, in turn creating greater equity in the amount of work done by both sexes (Clancy, Oparaocha, and Roehr 2004, 13).

Technical Solutions

As mentioned a few times, most technical problems require additional funding to be overcome. The more political support that is given in the form of grants, subsidies, and tax breaks, the quicker the technical barriers to realizing a transition away from fossil fuels will be overcome. Most technical issues, therefore, are more the result of political barriers to giving financial backing to renewable and alternative energy technologies. As discussed in Chapter 3, international cooperation has the potential to expedite technological breakthroughs and distribute the costs more fairly.

The problems that renewable energies have with localization, storage, and transmission can be alleviated with the use of hydrogen. In Chapter 1 we mentioned that hydrogen is actually an energy carrier, rather than an energy source. Because of its mobility, it can help all renewable energy sources reach their full potential. There is much consensus among experts that a renewable energy economy has to be coupled with the development of a hydrogen economy to meet the world energy demand. Furthermore, hydrogen emits no carbon dioxide, giving it great potential as a solution to the impending climate change crisis (Hoffman 2001, 17). With political support in the form of research and development grants, subsidies, and tax breaks, hydrogen has the potential to overcome any technical problems associated with it and become the main fuel for a renewable energy economy.

Policy Solutions

Now that the reasoning behind the push for an expedited transition to the use of renewable and alternative energy as a primary energy source has been reviewed, the next section presents the estimated prospects for renewable and alternative energy use as the main energy source in the United States.

Renewable and Alternative Energy Prospects in the United States

The United States consumes more energy than any other country in the world, over 21 percent of the annual world total of energy consumed, more energy than the entire continent of Europe consumes. By fuel type, the United States consumes almost 25 percent of the world's oil, exactly 22 percent of the world's natural gas, and almost 20 percent of the world's coal. The next largest energy consumer in the world is China, which consumes almost 16 percent of the world total of energy, with its high rates of growth expected to continue. In the meantime, these figures clearly reflect that the United States is by far the largest consumer of energy in the world (British Petroleum 2007, 40).

This copious consumption of fossil fuels has brought renewable and alternative energy to the forefront of the U.S. energy policy debate. Nuclear energy comprises just over 8 percent of the total energy consumed in the United States. While no new licenses have been granted since 1976, the Tennessee Valley Authority's Watts Bar reactor, which obtained its license in 1972, did not begin operating until 1996, making it the most recent reactor to be added to the electrical grid in the United States (Energy Information Administration 2007a). Although no new reactors have been licensed for construction since 1976 and none have come online since 1996, the U.S. Nuclear Regulatory Commission states that uprating (increases in electricity production by plants currently in operation) of operating reactors is expected to contribute over 4,000 megawatts of energy over the next several years. Thus far, uprating is the dominant policy for increasing electrical output from nuclear energy. The Energy Act of 2005, however, set new targets for increasing nuclear energy production, encouraging the construction of new reactors and creating a flurry of debate over the practicality of doing so. Despite lucrative tax breaks for new nuclear reactor construction, no U.S. companies have applied for a license to build a new reactor (Energy Information Administration 2007a).

The public is more supportive of expanding renewable energy capacity than expanding nuclear energy capacity. Currently, the United States produces 9.5 percent of its energy from renewable energy sources. Of that amount, almost half is produced from biomass, 42 percent from hydroelectric dams, 5 percent from geothermal, 3.7 percent from wind energy, and 1 percent from solar energy. Expected projections predict that, under current policies, renewable energy production will increase less than 2 percent by the mid-2020s (Energy Information Administration 2007a, 279).

Studies conducted by the U.S. National Renewable Energy Laboratory show that the potential for renewable and alternative energy development and production in the United States is large. Solar energy and geothermal potential is greatest in the Southwest and Greater Western area of the country. Wind power resources are located intermittently throughout the country but tend to be concentrated in the Great Plains states like North Dakota, South Dakota, and Kansas. Large wind resources are also available in the Southwest region. Biomass resources are more concentrated in the East and in parts of the Northwest. Wave power resources are located up and down both the East and West coasts of the country. OTEC energy, discussed in Chapter 1, has potential in the areas of the United States located between the latitudes of 20 degrees north and 20 degrees south. Its potential for producing electricity, hydrogen, and desalinized water makes investment into this technology appealing. The Northeast is the area that is the most lacking in renewable energy potential, yet it has a higher population density than other regions. For a complete switch to renewable energy, solutions to the problems of transmission and storage have to be fully developed. As discussed in Chapter 1, hydrogen as an energy carrier has the greatest potential to meet this challenge of renewable energy use. Although still very localized, the greatest hydrogen potential is located more centrally in the Great Plains states with less potential elsewhere the country.

While the potential exists for renewable and alternative energy to displace fossil fuels as the dominant energy source in the United States, it is unlikely that such a transition will occur without significant government intervention through a variety of policies to stimulate the technological development and market penetration of feasible technologies. Numerous correlations have been found between the growth of renewable energy production and policies that encourage such growth. It is particularly true that when financial incentives like tax credits and subsidies are revoked, the rate of growth of the affected technology slows accordingly. The next section gives an overview of the specific renewable energy policies that have been enacted in the United States to meet the prospects for energy production from renewable and alternative sources.

Renewable Energy Policies in the United States

Earlier we discussed the arguments held by those who think that government should do nothing and simply allow market forces to drive a transition to renewable energy. Proponents of government intervention argue that market forces may not promote an energy transition quickly enough to prevent serious environmental destruction and economic shocks due to the imminent decline of fossil fuel supplies and the political instability of oil-producing regions. Only through government intervention can the development of technology be encouraged and the timely adoptions of renewable and alternative energy become possible. Thus far, relying on market forces like supply and demand has not stimulated a significant transition to renewable energy technologies. Hence, this section and the following chapter focus on the policies that involve government intervention to promote renewable and alternative energy use.

In the United States, the majority of these policies are enacted at the state level rather than at the federal level. Market-based policies, which are explained in detail later, have been the predominant type of policy used by the U.S. federal government. States have been more aggressive in their passage of policies to promote renewable energy with the renewable portfolio standard growing in popularity as a policy mechanism to promote renewable energy development.

Market-based Policies.   The first type of policy is called a market-based policy because it works with the economic system to encourage competition of renewable and alternative energies with conventional energies. The goal of such policies is to create and support energy markets for renewable and alternative energy without hurting the economy. Ironically, these types of policies encourage market freedom with intervention that seeks to ensure that the market chooses renewable and alternative energy. While such policies certainly nudge the market in a particular direction, they allow for more flexibility in meeting certain goals than regulatory policies tend to. There are many manifestations of this type of policy.

Green Energy Market Development.   Green energy market development allows consumers to choose between green and brown energy. Green energy includes all renewable and alternative energy sources, and brown energy refers to energy derived from fossil fuels. Under this policy, companies are required to give consumers a choice between these two types of energy, and then they are required to produce as much electricity with green energy as is demanded by consumers. Currently, there is no federal policy to create a national green energy market because most energy issues devolve to the states. In 1993, California was the first state to implement this type of policy, which currently does not include nuclear energy. While this type of policy forces the market to give consumers the option to buy renewable and alternative energy, it does not guarantee that consumers will pay the extra costs of this type of energy, nor does it account for the lack of consumer knowledge about green energy options.

Centralized Bidding System.   Another type of market mechanism is a centralized bidding system combined with supply-side subsidies. In this policy, government allows renewable energy generators to bid on prices they are willing to provide energy for, and then the government makes regional electricity companies buy that energy. Unlike green energy market development policies, consumers do not have to pay more for this type of energy than they would for traditional energy sources because the government subsidizes the cost. Centralized bidding systems, in tandem with supply-side subsidies, not only encourage increased renewable and alternative energy use, but they reduce the price of such energy so that it is more desirable to consumers. The problem with this type of program is that it favors the cheaper sources of renewable and alternative energy, while providing no support for market penetration of the more expensive energy sources. Furthermore, renewable and alternative energy technologies that are not yet commercial, like ocean power and hydrogen, gain nothing from this type of policy because it is impossible for them to underbid already established energy sources like wind and solar. These policies tend to be more popular in Europe than the United States, though states like New York are beginning to implement them to meet their renewable energy portfolio standards, a type of policy that will be discussed below in the section on renewable energy portfolio standards.

Green Certificates.   Green certificates, also known as tradable renewable certificates (TRCs), green tags, and renewable obligation certificates (ROC), are becoming more prevalent as a solution for fostering a green energy market. Green certificates are granted for environmental attributes produced by green energy generators, which can then be traded on the secondary market. This allows renewable and alternative energy producers to sell the electricity they produce as well as the positive externalities of their product. For example, because renewable energy electricity sources emit much less carbon than traditional energy sources, producers could sell carbon credits in a carbon trading market. Because this policy is fairly new, its impact on the renewable and alternative energy market is difficult to assess. It is praised for its low cost of implementation and for causing an expected reduction in the cost of renewable and alternative energy. But, because green certificates are dependent on consumer demand to be successful, there is no guarantee that they will stimulate a significant increase in renewable and alternative energy production and use. The government merely allows these certificates to be sold by nongovernmental entities; it does not create any incentive for people to buy them. Green certificates currently exist in the United States, but there are no policies in effect to generate demand for them.

Net Metering.   Net metering is another type of policy created to work with the market. Net metering rules allow electricity customers to generate electricity from renewable energy and use it to offset electricity they would have bought at retail price from utility companies. Essentially, the energy meter rolls back when customers generate more electricity than they are using. This allows customers to create energy without having to deal with storing excesses for times when energy generation is low. This has the considerable advantage of reducing renewable energy investment because it eliminates the need for storage devices like batteries. Unfortunately, the initial costs of such an investment to individual consumers are still high with a long payback. The rate of individuals investing in personal renewable energy systems has not grown considerably despite the fact that net metering rules are used in 41 states to promote consumer investment in renewable energy.

Electricity Feed-in Laws.   A type of nonmarket-based policy is the so-called feed-in law, also known as an advanced renewable tariff (ART). Feed-in laws require utilities to pay higher prices for electricity produced by renewable and alternative energy companies in order to stimulate investment in renewable technologies. The extra costs are passed on to customers. This type of policy is generally effective at promoting renewable energy development, though not through the use of market mechanisms. Because the price for the renewable and alternative energy company's product is guaranteed, there is no incentive for them to reduce costs. This also prevents competition between renewable and alternative energy suppliers, resulting in excessive profits for producers. A related policy now being explored by a few states is the renewable tariff. Renewable tariffs are similar to feed-in laws, but they attempt to address the downfalls of feed-in laws by differentiating tariffs according to technology and by requiring a review of the renewable or alternative energy program to determine its robustness. Internationally, the renewable tariff is touted as one of the most successful policies for stimulating a rapid transition to renewable energy.

Renewable Energy Portfolio Standards. Another nonmarket-based policy mechanism that has been gaining popularity in the United States is the renewable energy portfolio standard (RPS). The RPS sets mandatory goals to specific actors, like utility companies, for renewable energy use and then lets the market determine the least expensive way to get there. The policy penalizes those who do not reach the goal. Most RPSs allow the trading of excess renewable energy production to actors who do not reach the goal. Under RPSs, a certain level of renewable energy production is guaranteed, the administration and bureaucratic costs for the government would be low, and price pressure would be maintained. However, this policy does not deal with the differing costs across technologies, nor is the goal set based on price or performance. Although RPSs allow the market to determine how to reach the goal, it does not ensure that the most efficient technologies will be used to get there because the high initial capital costs often discourage utility companies from choosing the best technology. Currently, 31 states and several cities have enacted RPSs, although not all of them are mandatory. There is no federal RPS. Texas has received considerable attention for its ambitious RPS program. The state is expected to meet the high standards that were set within the time period determined by the RPS, a total of 10,000 megawatts, or about 9 percent of the state's electricity demand, by the year 2025 (Rabe 2006, 11).

Other Policies. Other policies that are not directly geared toward stimulating market development include financial incentives like subsidies for research and development of renewable and alternative energy technologies and tax incentives for using such technologies. Although the federal government has enacted many policies dealing with renewable and alternative energy, the federal government tends to enact these types of policies the most. We now look specifically at some of the recent federal policies.

Current Federal Policies

The United States Department of Energy (DOE) is responsible for overseeing all federal programs and policies related to renewable and alternative energy. The Million Solar Roofs Initiative is one of the programs that the DOE currently oversees. Since its enactment in 1997, 300,000 homes and buildings have been outfitted with solar power energy systems. The goal is to have a million homes and buildings using solar power by 2010. The Wind Powering America Initiative, established in 1999, is a regionally based collaborative program that seeks to meet 5 percent of the nation's energy needs with wind energy by 2020 and to triple the amount of wind energy generated in states with a high wind capacity (to more than 20 megawatts total). Its main feature is the support of public–private partnerships in wind energy development. Within the DOE exists the Energy Information Administration (EIA). The EIA, a statistical agency of the DOE, collects objective data, issues forecasts and analyses for policy making, follows the efficient energy markets, and encourages public understanding of the interconnections between the energy market, the economy, and the environment.

A very important feature of the DOE is the National Renewable Energy Laboratory (NREL). Designated in 1991, the NREL is the principal laboratory for the U.S. Department of Energy's Office of Energy Efficiency and Renewable Energy. It is based in Golden, Colorado, with laboratories near Boulder and offices in Washington, D.C. The main purpose of the NREL is to drive technology from the "proof of concept" stage to the market penetration stage through research and applications of renewable energy technologies. Under the umbrella of the NREL organization is the National Center for Photovoltaics, the National Wind Technology Center, the National Bioenergy Center, and the Hydrogen Technologies and Systems Center. Underdeveloped technologies like ocean energy are undergoing development in conjunction with the private sector in the technology transfer office of the NREL.

The NREL is involved in the Solar America Initiative, which was enacted in 2006 under the auspices of the Advanced Energy Initiative. The Solar America Initiative is one of the widest ranging national policies geared toward market development for a renewable or alternative technology. It seeks to achieve market competitiveness for PV and CSP technologies by the year 2015. The initiative will first focus on research and the development of solar energy technologies, and then it will fund public–private partnerships to ensure market penetration of the technologies. It will also require that the federal government increase its use of solar energy by 5 percent by 2015.

Although these certainly contribute to an increase in the amount of renewable energy used, skeptics warn that they are not enough to result in a transition away from fossil fuels in time to prevent climate change or economic shocks from fuel shortages. The sentiments seem to be shared across the country, as states enact their own, more stringent policies to facilitate a speedy transition to renewable energy. We turn now to some examples of these policies.

Current State Policies

Besides the preceding policy types, one of the more popular state strategies for encouraging renewable energy development is to deregulate or restructure the energy market. Deregulation involves a transition from a market owned and regulated by the government to one that is an open competitive market. It is argued that true deregulation can never occur because government will always be required to intervene in some way or another. "Restructuring of the market" is therefore a more appropriate phrase. Restructuring still involves a degree of regulation, though it is encouraging the competition and privatization of the wholesale and retail electricity markets. Whatever it is called, it is a mix of government intervention and the increased use of markets to facilitate competition of energy prices. It is in such an environment that renewable energy must now compete. As of 2003, electricity markets are being actively restructured in 18 states, four states are facing delays in the process, and California has suspended the restructuring of its market.

Other policies that have been implemented in states across the nation include those aimed at building green energy markets. Contribution programs give consumers the option to round their energy bills up to the next dollar to stimulate renewable energy development. Wind programs offer electricity from wind for a surcharge. These programs have been very popular and have tended to sell out their capacity. Mixed fuel programs offer options for renewable energy from a variety of sources, including small hydro and methane capture from landfills.

While many of these policies have the potential to expedite the transition to renewable and alternative energy, they are not without their flaws. The drawbacks of the various types of programs have already been discussed. The last section of this chapter addresses some of the obstacles to successful renewable energy policy and the solutions to address them.

Policy Obstacles

As discussed in Chapter 1, technology emerges in stages. The stage that a technology is in has serious implications for what types of policies will work. While some argue that policies should focus on technologies that are ready for market penetration, some technologies are unlikely ever to reach this stage without supportive policies, especially those that fund research and development. Regardless, technologies that have not reached the breakthrough stage are going to require different policies than those that have in order for them to become competitive in the energy markets. Commercially viable technologies need policies that encourage market development in addition to research and development funding. Furthermore, there is no such thing as a one-size-fits-all policy. For government to aid in the increased acceptance of renewable energy, the optimal mix of policies for each type of technology must be achieved. Until all negative externalities of fossil fuels are fully accounted for (including ghg emissions, pollution, and all of the other environmental drawbacks of fossil fuel use), policy interventions are necessary for renewable and alternative energy to enter the energy market and become widely available enough for their use to offset the harms caused by fossil fuels.

The question then becomes what type of policy intervention should the government use? It is helpful to first look at some reasons why policies fail to achieve their objectives. Most experts agree that governments need to enact policies to encourage private sector investment into renewable energy projects. This can be done by distributing the costs of renewable energy widely to consumers and taxpayers, through policies that offer tax breaks to investors, low-cost loans to consumers, and subsidies for research. The issue of costs should not be the main focus of policy objectives, in accordance with the normative forecasting approach. To get the desired results, the focus should be on the outcome, not on the expense of getting there. In the case of global warming, it may not be affordable in the long run to cut costs in the short term.

Policy objectives, especially for policies that involve the private sector, should be well specified. If goals and targets are not specific, there is a good chance that the real purpose of enacting such a policy will not be met. An example of this is the restructuring of the energy markets. While the overall objectives are to create a competitive market where electricity costs are low, at the same time leading to the increased use of renewable energy, simply opening up the market does not guarantee that these goals will be reached. In the case of California, specifically, restructuring has been halted as the effects of it are being considered. It has been found that restructuring can actually raise the price of electricity, a phenomenon that is related not to specifying the role of the private sector in meeting the objectives of the policy. For utility companies, restructuring is a means for increasing profits, a goal that can often conflict with attempts to stimulate investment in expensive technologies.

Another reason that policies can fail to achieve their objective is the lack of market stability due to poorly developed policies. Incentives for investment in renewable energy will not achieve the objectives of the policy if private sector entities do not trust the government to protect their interests. Policies that have caps on their incentives will stimulate investment or production only up to the point where the incentives stop. Placing time frames, or sunsets, on policies meant to encourage long-term investment is counterproductive as well. If the private sector is expected to make long-term investments only when incentives are provided, then it makes little sense to remove the incentives before investments have been recouped. An example of an unstable policy is Wind Energy Tax Credit, which has to be reviewed and extended every few years (Komor 2004). Such a review would be a viable policy if it was based on performance rather than on budget concerns. A performance-based review would not only stabilize policy, it would create incentives for the private sector to meet the targets of the policy.

The rejection of renewable energy projects due to costs and lack of public acceptance are other impediments to successful policy. Such barriers can be overcome through the involvement of stakeholders in policy formulation and especially through public ownership of renewable and alternative energy developments. As shown in Chapter 3, public acceptance has grown in European countries where locals are given a stake in the development. Furthermore, policies and developments that have not involved all stakeholders affected have failed.

The economic burdens of a speedy transition to renewable and alternative energy should be dispersed as widely as possible, especially since the benefits will be felt on a large scale once the transition is complete. Benefits and profits should also be distributed appropriately, preventing excessive and inefficient gains by companies as demonstrated by feed-in laws. This may need to be done on an international scale with agreements like the Kyoto Protocol, which seeks to alleviate environmental damage from energy use on a proportional basis. Such agreements can also address another policy issue, which is technology development and transfer. If the burden of getting technology through the various stages of development is shared globally, the solutions could become commercially viable much sooner. We turn now to a discussion of the global and regional agreements in place to provide these solutions.

Global Perspectives

Over the past 50 years, it has come to the attention of governments around the world that energy problems are often global in nature. The problems are economic, environmental, and political, and they are directly related to the increasing demand and consumption of fossil fuels, as already discussed. Oil price shocks have been proven to affect the well-being of the entire global economy. Climate change has been significantly attributed to the global use of fossil fuels by a consortium of scientists from nations around the world (International Panel on Climate Change 2007b, 2). Satellites have discovered that air pollution is becoming transcontinental, even traveling across the Pacific Ocean (Holloway, Fiore, and Galanter-Hastings 2003). International conflicts have been attributed to dependence on imported oil. As a result, solutions to these problems also often have to be global in nature. This section examines the major multilateral initiatives that have been designed in response to energy-related problems. These agreements either directly encourage a transition to renewable and alternative energy use or have goals that will indirectly result in an increase in production from these sources. Often they are in addition to many of the participating country's domestic renewable and alternative energy policies. In any case, they are expected to be a crucial part of the solution.

Kyoto Protocol

The Kyoto Protocol, created in 1997, is an amendment to the international agreement, the United Nations Framework Convention on Climate Change, signed by several countries to reduce ghg emissions like carbon dioxide. As of the publication of this book, the United States has not ratified the protocol, making its success in meeting its ultimate goal of combating climate change uncertain. While the main purpose of this agreement is to combat climate change using a variety of policy mechanisms, a transition to renewable energy will have to be part of the solution because fossil fuel emissions are the main cause of human contributions to the increase of greenhouse gases in the atmosphere. The main policy mechanism of the Kyoto Protocol is a carbon cap and trade system. The system sets allowable carbon dioxide emissions to 1995 levels. This is a flexible market–based approach, which would allow some countries to produce more emissions as long as they have acquired permits from other countries that are producing less. Such an approach does not mandate an increase in renewable energy use per se, but, to meet the commitments set in the agreement, the production of energy via renewable sources will increase as a matter of economic sensibility. As indicated in an EIA study on the Kyoto Protocol, renewable energy will increase to meet emission goals in sectors where it is cheapest to switch from reliance on fossil fuels to reliance on renewable energy. The increase in renewable energy production, however, would not be significant (Energy Information Administration 1998, xx). Perhaps the main reason for the Kyoto Protocol's lack of impact on an energy transition is related mostly to the fact that it only requires a 5 percent reduction in current carbon dioxide levels. This amount is not only inadequate to reach the ultimate goal of halting or reducing climate change, it is small enough that it can be reached without major shifts in energy sources. While the Kyoto Protocol will result in some increases in renewable energy production throughout the world, countries will have to voluntarily go beyond the 5 percent target for the agreement to have a significant effect on the use of renewable energy, as shown in the section on the European Union.

The Asia Pacific Partnership on Clean Development and Climate

The Asia Pacific Partnership on Clean Development and Climate, also known as A6, was enacted in 2006. It is a nonbinding negotiation among the United States, Australia, India, Japan, China, and South Korea to facilitate the cooperation and transfer of technology that will help reduce ghg emissions. Because it is nonbinding, there are no mandatory limits on emissions as in the Kyoto Protocol. One of the main objectives of the group is to work on overcoming barriers to the adoption of renewable energy technologies. The main concern of the task force on Renewable Energy and Distributed Generation is to promote the use of

the most advanced technologies or those that have already pene-
trated the market like solar, hydro, geothermal, tidal, and alternative fuels like ethanol (Asia Pacific Partnership on Clean
Development and Climate 2007, 3). The goals set in the A6 Renewable Energy and Distributed Generation Task Force Action
Plan are vague and voluntary. There is no time frame for meeting
these goals, nor does the plan seek to generate a set amount of energy from renewable energy sources. While such an agreement is
better than no agreement at all, critics argue that, without en-
forcement of mandatory targets, the agreement will do little to
promote substantial renewable energy use or reduce ghg emissions (*Economist* 2006, 46).

Johannesburg Renewable Energy Coalition
The Johannesburg Renewable Energy Coalition (JREC) was
signed in Johannesburg, South Africa, in 2002 at the World
Summit for Sustainable Development (WSSD). As of 2004, 88
countries from the European Union (EU), the Middle East,
Africa, the Caribbean, and the Pacific Islands belong to the
coalition. Countries with the highest energy consumption rates
like the United States, Australia, China, and India are not members. The function of the coalition is similar to that of the A6
partnership, which is to facilitate cooperation and coordination
among member countries in an effort to specifically increase
the dissemination of renewable energy technologies. Unlike
A6, however, JREC requires members to adopt targets for renewable energy production. The main purpose of JREC is to
achieve the adoption of these targets by its international members at the 15th session of the United Nations Commission for
Sustainable Development in 2007. While the 15th session did
result in many positive affirmations of the need for global renewable energy development, no binding time tables were set
for when targets should be met. As discussed, the less government intervenes to encourage renewable energy development,
the slower this development will occur because market forces
are driving fossil fuel development.

Conclusion

This chapter has revealed the underlying scientific, environmental, political, and social concerns that have surfaced over

the last few decades, contributing to the debate over the use of renewable and alternative energy as a solution for these concerns. There is a general consensus about issues such as global warming that the precautionary principle should be employed and that government intervention is necessary to meet this end. We have provided a review of the controversies surrounding a transition to renewable and alternative energy sources. Next we explored the broad advantages associated with renewable and alternative energy use, followed by a look at the actual prospect and potential of renewable energy in the United States to replace fossil fuels as the dominant fuel. Following that assessment was a look at the types of federal and state policies in place to promote renewable and alternative energy. Then we discussed solutions and suggestions for policy development to overcome the challenges and barriers of current renewable energy legislation. Finally, we reviewed some major global and regional efforts to provide solutions

Chapter 3 provides a similar discussion to that in Chapter 2, but in the context of the various countries. The prospects for renewable and alternative energy as a viable solution are explored further. By examining policies that have been enacted and policies that have failed internationally, we provide further perspectives on how to achieve a global transition to renewable and alternative energy.

References

Asia Pacific Partnership on Clean Development and Climate. "Renewable Energy and Distributed Generation Task Force Action Plan." November 2007. www.asiapacificpartnership.org/RenewableEnergyTF.htm

British Petroleum. *Statistical Review of World Energy Full Report 2007.* June 2007. www.bp.com/liveassets/bp_internet/globalbp/globalbp_uk_english/reports_and_publications/statistical_energy_review_2007/STAGING/local_assets/downloads/pdf/statistical_review_of_world_energy_full_report_2007.pdf

Buddemeier, Robert W., Joan A. Kleypas, and Richard B. Aronson. *Coral Reefs and Global Climate Change: Potential Contributions of Climate Change to Stresses on Coral Reef Ecosystems* (report from the Pew Center on Global Climate Change). Arlington, VA: Pew Center on Global Climate Change, 2004. www.pewclimate.org/global-warming-in-depth/all_reports/coral_reefs

Campbell, C. J. *The Coming Oil Crisis.* Brentwood, UK: Multi-Science Publishing and Petroconsultants, 1997.

Center for American Progress. *American Energy: The Renewable Path to Energy Security* (report prepared for the Worldwatch Institute). Washington, DC: Worldwatch Institute, 2006. americanenergynow.org/

Chambers, A. "Wind Power Spins into Contention." *Power Engineering* 104 2 (2000): 14–18.

Champion, Marc. "Glitch Forces Georgia to Pay up for Russian Gas." *The Wall Street Journal* (December 23, 2006) 4.

Clancy, Joy, Sheila Oparaocha, and Ulrike Roehr. "Gender Equity and Renewable Energies" (thematic background paper). Bonn: International Conference for Renewable Energies, 2004.

Database of State Incentives for Renewables and Efficiency (DSIRE). "Renewables Portfolio Standard." 2007. www.dsireusa.org/library/includes/incentivesearch.cfm?Incentive_Code=TX03R&Search=Type&type=RPS&CurrentPageID=2&EE=1&RE=1

Deffeyes, Kenneth S. *Beyond Oil: The View from Hubbert's Peak.* New York: Hill and Wang, 2005.

Economist. "More Hot Air; Asian Environmentalism." *Economist* 378, 8460 (January 15, 2006): 46.

Energy Information Administration (EIA). Chapter 7, "Environment and Renewable," in *Energy in Africa.* 1999. www.eia.doe.gov/emeu/cabs/chapter7.html#maec

Energy Information Administration (EIA). "International Energy Annual (IEA) 2005." 2006. www.eia.doe.gov/emeu/iea/contents.html

Energy Information Administration (EIA). "Annual Energy Review (AER) 2006." Report DOE/EIA-0384(2006). 2007a. www.eia.doe.gov/emeu/aer/overview.html

Energy Information Administration (EIA). "International Energy Outlook 2007." Report DOE/EIA-0484(2007). 2007b. www.eia.doe.gov/oiaf/ieo/index.html

"Global Warning; the World's Scientists Agree, Again, That Climate Change Is a Big Problem." *Washington Post* (February 5, 2007). Lexis-Nexis, www.lexisnexis.com

Goodstein, David. *Out of Gas: The End of the Age of Oil.* New York: W. W. Norton, 2004.

Hammerschlag, Roel. "Ethanol's Energy Return on Investment: A Survey of Literature from 1990–Present. *Environmental Science and Technology* 40 6 (2006): 1744–1750.

Hoffman, P. *Tomorrow's Energy: Hydrogen, Fuel Cells, and the Prospects for a Cleaner Planet*. Cambridge, MA: MIT Press, 2001.

Holloway, T., A. Fiore, and M. Galanter-Hastings. "Intercontinental Transport of Air Pollution: Will Emerging Science Lead to a New Hemispheric Treaty?" *Environmental Science and Technology* 37, 20 (2003): 4535–4542.

Huber, Peter W., and Mark P. Mills. *The Bottomless Well: The Twilight of Fuel, the Virtue of Waste and Why We Will Never Run Out of Energy*. New York: Basic Books, 2005.

International Energy Agency (IEA). *World Energy Outlook 2006*. 2006. www.worldenergyoutlook.org

International Panel on Climate Change (IPCC). "Summary for Policy Makers." *Climate Change 2007: The Physical Science Basis*. February 2007a. www.ipcc.ch/pdf/assessment-report/ar4/wg1/ar4-wg1-spm.pdf

International Panel on Climate Change (IPCC). "Summary for Policy Makers." *Climate Change 2007: Impacts, Adaptation and Vulnerability*." April 2007b. www.ipcc.ch/pdf/assessment-report/ar4/wg2/ar4-wg2-spm.pdf

Jameson, Angela. 2005. "Uranium Shortage Poses Threat." *TimesOnline*. Business Section. August 15. business.timesonline.co.uk/tol/business/industry_sectors/industrials/article555314.ece

Kennedy, Robert F. Jr. "An Ill Wind off Cape Cod." *The New York Times*. Editorials Section. December 16, 2005. www.nytimes.com/2005/12/16/opinion/16kennedy.html

Khilyuk, L. F., and G. V. Chilingar. "On Global Forces of Nature Driving the Earth's Climate. Are Humans Involved?" *Environmental Geology* 50, 6 (August 2006): 899–910.

Komor, Paul. *Renewable Energy Policy*. Lincoln, NE: iUniverse, 2004.

Laherrere, Jean. "Oil and Gas: What Future?" Paper published for the Groningen Annual Energy Convention. November 21, 2006. www.oilcrisis.com/laherrere/groningen.pdf

Leggett, Jeremy. *The Empty Tank: Oil, Gas, Hot Air, and the Coming Global Financial Catastrophe*. New York: Random House, 2005.

Lindzen, Richard S. "Is There a Basis for Global Warming Alarm?" The Independent Institute. October 21, 2005. www.independent.org/publications/article.asp?id=1714

MacGill, I., M. Watt, and R. Passey. 2002. "The Economic Development Potential and Job Creation Potential of Renewable Energy: Australian Case Studies." Report Commissioned by Australian Cooperative Research Centre for Renewable Energy Policy Group, Australian Ecogeneration Association and Renewable Energy Generators Association. 2002.

Michaels, Patrick J. "Posturing and Reality on Warming." *Washington Times* (October 16, 2003) A17.

Myers, Norman, and Jennifer Kent. *Perverse Subsidies: How Tax Dollars Can Undercut the Economy and the Environment.* Washington DC: Island Press, 2001.

Odell, P. R. *Fossil Fuel Resources in the 21st Century.* London: Financial Times Energy, 1999.

Perlin, John. "Solar Power: The Slow Revolution." *Invention and Technology Magazine* (Summer 2002). www.americanheritage.com/articles/magazine/it/2002/1/2002_1_20.shtml

Piven, Frances Fox. *The War at Home: The Domestic Cost of Bush's Militarism.* New York: New Press, 2004.

Rabe, Barry G. 2006. *Race to the Top: The Expanding Role of U.S. State Renewable Energy Portfolio Standards* (report for the Pew Center on Global Climate Change). Arlington, VA: Pew Center on Global Climate Change, 2006. www.pewclimate.org/global-warming-in-depth/all_reports/race_to_the_top

Revkin, Andrew C. "Poorest Nations Will Bear Brunt as World Warms." *The New York Times* (April 1, 2007) 1.

Riley, Peter. *Nuclear Waste: Law, Policy and Pragmatism.* Burlington, VT: Ashgate Publishing, 2004.

Saddler, Hugh, Mark Disendorf, and Richand Denniss. "A Clean Energy Future for Australia." WWF International. March 2004. wwf.org.au/ourwork/climatechange/cleanenergyfuture/

Smil, Vaclav. *Energy at the Crossroads: Global Perspectives and Uncertainties.* Cambridge, MA: MIT Press, 2003.

Stroeve, J., M. M. Holland, W. Meier, T. Scambos, and M. Serreze. "Arctic Sea Ice Decline: Faster than Forecast." *Geophysical Research Absracts* 34 (2004): L09501, doi:10.1029/2007GL029703. www.agu.org/pubs/crossref/2007 . . . /2007GL029703.shtml

Swisher, Randall, and Kevin Porter. "Renewable Policy Lessons from the U.S.: The Need for Consistent and Stable Policies." In Karl Mallon, ed. *Renewable Energy Policy and Politics: A Handbook for Decision Making.* London: Earthscan, 2006, 185–198.

Tennekes, Hendrik. "A Skeptical View of Climate Models." 2007. www.sepp.org/Archive/NewSEPP/Climate%20models-Tennekes.htm

United Nations. *Agenda 21: The UN Programme of Action from Rio.* New York: United Nations, 1992.

U.S. Government Accountability Office. "Venezuela Report Summary." 2006. www.lugar.senate.gov/energy/venezuela/pdf/GAO_Report_Venezuela_Summary.pdf

U.S. Nuclear Regulatory Commission. "Fact Sheet on Power Uprates for Nuclear Plants." 2008. www.nrc.gov/reading-rm/doc-collections/fact-sheets/power-uprates.html

3

Worldwide Perspective

Introduction

This chapter focuses on the state of renewable and alternative energy use and policy throughout the world. Specifically, it looks at developed and developing countries in the light of important renewable and alternative energy issues. Each of the two types of nations, it will be shown, faces different barriers for the successful advancement of renewable energy technologies. Like the United States, most developed countries of the world face economic barriers to and concerns about economic efficiency. With the exception of China and India, developing and underdeveloped countries face much more than economic barriers. Although they tend to be more concerned about their economies than the environment, a lack of funding for the high initial costs of renewable energy investment is not the only issue preventing development of the renewable energy sector. Developing countries are at a disadvantage due to a lack of technical expertise in the field of renewable energy. Also, they often are missing the governmental support necessary for developing and imposing regulations required to build a renewable energy sector. In some countries, the fossil fuel industries have enough of a say in politics that they are able to block policies that would increase renewable energy production.

To overcome these problems, developing countries need assistance from developed countries and the international community. Cooperative efforts have already begun to increase renewable energy development. Fears over nuclear weapon proliferation and accidents associated with the use of nuclear power have caused developed countries to resist giving developing countries help with nuclear power development. In some cases, developing countries seeking to develop nuclear energy are being discouraged by developed countries, at times creating hostile relationships. The need for international energy policies is increasing every year.

The first section of this chapter addresses policies adopted by the European Union (EU) as a whole and national policies adopted by EU member states. The following section discusses Asia-Oceania, paying particular attention to India, China, Japan, and Australia. The next section focuses on countries in North and Central America. The last section addresses South America.

Europe

As of 2007, 27 countries on the continent of Europe have joined the European Union. The EU is a political and economic intergovernmental organization. The European Commission (EC) is the executive body that oversees many of the policies being jointly implemented by EU member countries. All of Europe consumes 19 percent of the total amount of world energy produced, and Europe produces less than two-thirds of the total energy it consumes. As a result, the continent, as a whole, must import more than one-third of the energy it consumes (Energy Information Administration 2006, 302). Of this amount, Germany consumes the most energy, followed by France, the United Kingdom, and Italy, respectively (Energy Information Administration 2006, 303). Because of its high energy use and expected growth in consumption, the EU has taken the role as the world leader in adopting many policies on climate change and renewable and alternative energies. These policies often have binding targets on carbon dioxide emissions reductions and renewable energy production for member states. Individual countries enact national policies to meet the targets laid out in EU directives. This section first gives an overview of the policies enacted by the EU member states as a whole, followed by a discussion of the specific national policies of individual member states to meet the renewable and alternative energy goals of the EU directives.

EU Renewable and Alternative Energy Policies

In 1997, to help meet the requirements of the Kyoto Protocol, the EC set out its strategies and action plan in the European Union white paper. The white paper, called "Energy for the Future: Renewable Sources of Energy 1997," set a goal to double the amount of renewable energy used by EU member countries to 12 percent. To do this, the paper created several legislative proposals to reduce the use of fossil fuel for electricity, transportation, and heating. Its main policy is the Renewables Directive, which requires that 22 percent of total electricity come from renewable energy sources by 2010. Another major piece of legislation falling under the white paper is a biofuels directive, which requires biofuels to provide 5.75 percent of the total fuels used for transportation. Also falling under the white paper legislative proposals is the Directive on Energy Performance in Buildings. This directive requires energy efficiency to be improved in the construction of new buildings.

More recently, the EC developed the Renewable Energy Roadmap in 2007, which provided a "long term vision for renewable energy sources in the EU" (Commission of the European Communities 2007a, 3). The Roadmap reported that the EU would fall short of its goal of getting 12 percent of its energy from renewable energy sources. This is because many of the national policies of individual countries are not good enough to reach the goals of the white paper. Despite this setback, the Roadmap established an even higher target for renewable energy. By 2020, renewable energy must account for 20 percent of total energy use in Europe. While the cost is expected to be very high, the Roadmap points out that similar investments were made for the coal and oil industries in the past. In March 2007, European leaders agreed to the target. The goal is related to another goal of reducing greenhouse gas (ghg) emissions by 20 percent by 2020 (Euractiv 2007b). It is important to note that these EU commitments are to renewable energy only, not to alternative energies like nuclear power. EU countries are divided on nuclear energy. France is the biggest supporter of expanding nuclear energy (Euractiv 2007a). In 2007, EU member states updated their national policies to meet the goals of the Renewable Energy Roadmap.

Directive on the Promotion of Electricity from
Renewable Energy Sources in the Internal Electricity Market

Also known as the Renewables Directive, the Directive on the Promotion of Electricity from Renewable Energy Sources in the

Internal Electricity Market was adopted in 2001 by the leaders of EU member countries. It requires that all EU states meet their electricity needs with at least 22 percent coming from renewable energy by 2020. As of 2005, electricity from renewable energy sources was at 15 percent. This amount is expected to increase to 19 percent by 2020. The major source for this increase is large-scale hydro, though electricity from other renewable resources is expected to continue growing in the future (Commission of the European Communities 2007b, 3).

Directive on the Promotion of the Use of Biofuels for Transport

In 2003, the EU adopted the Directive on the Promotion of the Use of Biofuels for Transport. The Biofuels Directive set a target for biofuels to provide 2 percent of the EU's fuel needs by 2005 and 5.7 percent by 2010. A progress report published in 2007 found that biofuels accounted for only 1 percent in 2005 and that it was unlikely that it would reach more than 4.2 percent by 2010. Again, the national policies of individual countries are not keeping up with the targets adopted in the white paper (Commission of the European Communities 2007, 6).

Electricity Feed-in Laws

Chapter 2 presented electricity feed-in laws, which require utilities to pay higher prices for electricity produced by renewable and alternative energy companies. Electricity feed-in laws are known as advanced renewable tariffs (ARTs) in Europe. ARTs are the most popular approach to encouraging renewable energy in EU member countries. To date, 19 EU member countries are using ARTs, which have been called the "world's most successful policy mechanism for stimulating rapid development of renewable energy" (Sijm 2002, 5). ARTs have seen growing success in EU countries like Spain, Germany, and Denmark. In these cases, ARTs resulted in over 40 percent of growth in wind power. As stated in Chapter 2, criticisms of ARTs are that they are not economical and that they result in higher prices than if such energies had developed in a competitive market.

Germany

Germany is the highest consumer of energy in the EU, although consuming only 3.1 percent of total energy worldwide (British Petroleum 2007, 37). In 2006, renewable energy made up 5.8 percent of the total energy and 12 percent of electricity generation in Germany (Federal Ministry of Economics and Technology 2007). In 2000, Germany became the first economically developed country to announce a phase-out of nuclear power. Nuclear power accounts for 27 percent of its electricity generation. Without nuclear power, Germany has had to pass stronger policies for renewable energy development than other EU countries. As a result, it is more successful at encouraging renewable energy (Runci 2005, 2). The major strategy of these policies has been a shift from publicly funded research and development (R&D) to policies that promote market penetration of renewable energy technologies.

As mentioned, electricity feed-in laws resulted in large increases in renewable energy production. Since 1990, when Germany's Electricity Feed-in Law was passed, renewable energy production has grown by 500 percent for biomass, 2,000 percent for wind, and 15,000 percent for solar photovoltaics (PVs) (Runci 2005, 1). This rise has made Germany one of the top renewable energy producers in the world (Runci 2005, 2). However, because there was hardly any development to begin with, a 15,000 percent increase sounds much higher than it really is.

Another important factor in the rise in renewable energy production in Germany was the passage of the Renewable Energy Law (EEG) in 2000. The goal of the EEG was to increase electricity produced from renewable energy by 12.5 percent by 2010. Like the electricity feed-in law, the EEG required that electricity producers buy renewably generated electricity at set prices. Wind energy has benefited the most from these policies, but solar continues to have difficulty competing in the market.

An Ecological Reform Tax passed in 1999 has contributed to a growth in biofuels by exempting it from taxes. A 100,000 Solar Roofs Program, passed in 1999, resulted in a 700 percent increase in installed solar PV capacity from 1998 to 2003 with the help of considerable government subsidies (Runci 2005, 10). Also passed in 1999 was the Market Incentive Program (MAP), which provided over €200 million in grants in 2003 for renewable energy systems (Runci 2005, 10).

Despite the growing success of Germany's renewable energy policies, obstacles must still be overcome. Many renewable energy policies, especially subsidies, have met resistance from competing interests like the coal industry, resulting in a reduction in subsidies in 2003 (Runci 2005, 16). Following the trend of general oppositions to renewable energy, Germany has also experienced public opposition to wind turbines. Experts say that, because Germany has produced so much power from renewables, it may be more cost-effective to import renewably generated electricity from other EU countries. The long-term future success of renewable energy production depends on whether the German government passes the right policies. However, it will be difficult for Germany to phase out nuclear power if renewable energy does not continue to grow at the rate it is growing now.

France

France is the second highest consumer of energy in the EU and consumes about 2.5 percent of worldwide total primary energy (British Petroleum 2007, 37). As of 2005, 6 percent of France's primary energy came from renewable energy, the majority of which came from large-scale hydro power plants (European Commission 2006b, 2.6.9). Nuclear energy provides over 70 percent of France's electricity needs (European Commission 2006b, 2.6.9). France is aiming to have 21 percent of its electricity come from renewable energy by 2010. National targets for biofuels have been set at 10 percent by 2015 (European Commission 2007a, 1). As the nuclear power percentage indicates, French society is much more accepting of the option of meeting its energy needs with nuclear power production. Consequently, France has been the main objector in EU energy policy discussions when other states propose that nuclear power should not count toward green energy targets. Although France has agreed to the energy targets set by the EU, it continues to call for the incorporation of nuclear energy as a solution for global warming (Cook 2007).

The General Directorate for Energy and Raw Materials (DGEMP) within the Ministry of Economy, Finance, and Industry is responsible for developing energy policy in France. France relies on two main policy mechanisms to promote renewable energy production. First, like Germany, France has experienced increases in renewable energy production, particularly wind, through ARTs. In addition to ARTs, calls for tenders have been issued. This means that the French government is soliciting the private sector to invest in large-scale renewable energy projects. Project Renaissance of Lyons is a good example of a tendering project. The purpose of the project is to reach specific goals for renewable energy and energy efficiency and to promote sustainability in the private construction sector. Through the financing of high overhead costs, Project Renaissance also seeks to encourage developers to invest in the renewable energy sector.

As of 2007, France has been unable to meet its goals for renewable energy electricity production and biofuel use. It is unlikely that France will meet its commitments for ghg reductions under the Kyoto Protocol (Commission of the European Communities 2007b). Renewable energy production fell from 15 percent in 1997 to less than 13 percent in 2004 (European Commission 2007a, 1). A biofuels target of 2 percent was not met in 2005, the country falling 50 percent short of this goal (European Commission 2007a, 1). It is unknown at this time whether France will proceed to meet its targets for renewable energy sources or continue developing additional nuclear power capacity.

United Kingdom

The United Kingdom (UK) is the third highest consumer of primary energy in the EU but consumes only about 2 percent of total worldwide energy (British Petroleum 2007, 40). In 2005, renewable energy provided just under 2 percent of the United Kingdom's total primary energy needs and almost 5 percent of its electricity (DTI 2006, 29). Nuclear energy accounted for 7.5 percent of the total energy in the UK (British Petroleum 2007, 40). The UK has implemented several national policies to meet the challenging goals put forth in the EU directives. The Energy Act of 2004 reflects the commitments of the UK under the white paper and requires that 10 percent of all energy come from renewable sources by 2010. The act also set up the Nuclear Decommissioning Authority to address shutting down and cleaning up nuclear energy sites. In addition, it created the British Electricity Trading and Transmission Arrangements (BETTA), which created a single wholesale electricity market that establishes a single electrical grid and will promote competition in the market.

In April 2002, the UK began encouraging renewable energy production with the Renewables Obligation. The Renewables Obligation requires that 10 percent of electricity must come from renewable sources by 2010, as long as the costs are not too high for consumers. The policy works like a renewable portfolio standard, as discussed in Chapter 2. In addition to the 10 percent target, the policy gives renewable obligation certificates (ROCs) to companies that produce more than 10 percent of their electricity from renewable energy. ROCs are similar to the green certificates discussed in Chapter 2. The ROCs create a market for renewable energy credits so that companies that are unable to meet the obligation can buy out of it by purchasing certificates from companies that can meet more than the 10 percent requirement.

Experts expect that renewable energy will contribute a total of 9.9 percent of electricity production by 2010, just under the target of 10 percent (National Audit Office 2005, 2). The UK government has identified barriers to the success of the renewables obligation. These include the lack of uniformity of planning systems across the country, upgrades needed for the grid network, unstable wholesale electricity prices, unstable policy frameworks, and the lack of commercial viability of some renewable energy technologies (National Audit Office 2005, 2–4). The result has been a dedicated response by the UK federal government to develop solutions to overcome these barriers.

Other Notable Policies of EU Member States

Sweden is a unique case when it comes to renewable energy. In 2005, Sweden became the first country in the world to pledge a commitment to fossil fuel independence by the year 2020 (Ministry of Enterprise, Energy and Communications 2007). Most of Sweden's renewable energy share comes from hydropower and more recently from biomass. Because of a reliance on biomass for over 60 percent of domestic heating needs, Sweden has increased its share of renewable energy use to 28 percent of the total energy supply (Ministry of Enterprise, Energy and Communications 2007). All other renewable energies account for less than 1 percent of the total energy supply. A recent policy aimed at increasing energy efficiency in buildings has a target of using 50 percent less energy by 2050. More specific programs for meeting the ambitious goal of fossil fuel energy independence by 2020 are still being developed by the newly created Commission on Oil Independence.

Denmark succeeded in eliminating the need to import energy from other countries, making it energy independent. Denmark is the world leader in offshore wind energy and is expected to meet its Kyoto Protocol emission reductions of 29 percent three to four years early (Commission of the European Communities 2007b, 7). Already, Denmark produces over 20 percent of its electricity from renewable energy sources and has ambitions to continue on the path toward a renewable energy economy (Danish Energy Authority 2005).

Spain has committed to a renewable energy target of 12 percent of overall energy and 30 percent of electricity by 2010 (International Energy Association 2007). To reach this goal, Spain is looking to increase energy production from solar, wind, and biofuels with 77 percent of financing for the projects coming from the energy industry (International Energy Association 2007). Spain is the second largest producer of wind energy in the world (Commission of the European Communities 2007b, 8).

The Netherlands has found success with electricity feed-in laws like France and Germany, but with biofuels rather than wind (Commission of the European Communities 2007b, 8). To meet its ghg emission reduction commitments under the Kyoto Protocol, the Netherlands has committed to a renewable energy target of 5 percent by 2010 and 10 percent by 2020. It is expected that it will reach these targets, although there have been reductions in financial support for large-scale biomass and offshore wind power (European Commission 2007b, 1).

These countries have the most successful policies for promoting renewable energy in the EU. As discussed, the electricity feed-in laws or ARTs are responsible for most of the increases in production of renewable energy because they require utilities to purchase the renewable energy produced by wholesale generators. Although it is argued that this policy mechanism is not the most economically efficient, it cannot be denied that it gets results. The mandatory purchase of electricity produced from renewable energy aids in the process of making these technologies competitive with fossil fuel sources. As a result, other, more market-friendly policies can be enacted with more potential for future success.

Because the EU drive to adopt renewable energy is the result of its commitment to lead the global effort to reduce ghg emissions under the Kyoto Protocol, the EU and its member states have more aggressive policies than the United States and other countries for promoting a transition to renewable energy. The general dislike by EU member states of nuclear power, with the exception of France and a few countries from Eastern Europe, reflects shared sentiments with U.S. society on the dangers of nuclear accidents and nuclear waste. Recent reports indicate, however, that nuclear power may have to become accepted as a means for reaching binding targets on emission reductions in EU countries (Ames 2007).

Although not a EU member state, Iceland has one of the most advanced renewable energy programs in the world. It is the only Western country to rely completely on renewable energy for electricity production. Geothermal supplies 83 percent of electricity, and hydropower supplies 17 percent. The country imports oil for transportation and coal for industrial uses, and it has no nuclear power plants in operation. Iceland is also one of two countries that have a real potential to produce hydrogen solely from renewable energy.

We turn now to a review of renewable and alternative energy policies in Asia and Oceania.

Asia and Oceania

Many aspects of the Asian energy situation make this region an important area for study. The first and most obvious is that the increasing population, combined with the growing economy in countries like China and India, is expected to create remarkable demands for energy, resulting in high rates of consumption. Energy forecasters have hardly begun to predict the effects on the energy supplies and markets if these demands are met with continued reliance on fossil fuels. Oil shortages and high energy prices are among a couple of the growing concerns that have prompted many Asian governments to begin passing policies to encourage renewable and nuclear power energy production. Asian mega cities are at the top of the list for the worst air pollution in the world, causing severe health problems to the people who live there. The pollution is due to the growth of industrial development and reliance on coal. Development in Asian economies is expected to contribute greatly in the near future to the global warming effect, causing worldwide attention to Asian energy issues. It is expected that China will surpass the United States as the top carbon emitter by 2008 (International Energy Association 2006). As a region, Asia consumes 31 percent of the world's primary energy, which is almost as much as North and South America combined (Energy Information Administration 2006, 302).

As discussed in Chapter 2, the A6 agreement and Kyoto Protocol do not set binding limits on ghg emissions or targets for renewable energy adoption for most Asian countries. Many of the national policies of Asian countries to encourage renewable and alternative energy production are concerned first and foremost with the benefits the technology can offer to the economy rather than reducing environmental impacts. Standards of living of rural residents are another concern; in many cases it is more economically efficient to develop renewable energy in these areas than to transport electricity by bringing in electrical lines from fossil fuel utility plants in urban areas. Because of these issues, renewable energy policies seldom impose binding targets.

While most countries in the West have halted nuclear energy development or are phasing it out altogether, the growing energy needs of Asia make it an attractive source of energy there. As international pressures to reduce ghg emissions increase, Asian countries are turning to nuclear energy development as a clean and efficient way to provide electricity for its growing populations. The Energy Information Association predicts that there will be considerable growth in the nuclear power industry in Asia as the U.S. and European markets are dwindling (Energy Information Administration 2002).

In our discussion of Asia, we turn first to the most rapidly growing country, China. The next section identifies the important renewable and alternative energy issues in India. Following that is a discussion of Japan and Australia. We then provide a section highlighting the other most notable renewable and alternative energy developments in the region.

China

As one of the fastest growing countries in the world, China ranks second in the world for energy consumption. With its rate of economic and population growth, it is expected to maintain that position into the coming decades (International Energy Association 2006, 47). The primary source of energy is coal, followed by oil and natural gas (European Commission 2006, 2.8.2). But as energy shortages loom, oil prices rise, and the environmental impacts of fossil fuel use are felt, the Chinese government has started taking action. A growing middle class has contributed to the rise in gasoline consumption as the number of automobile owners increases, resulting in gas shortages in some areas of the country. The rising price of oil and gas has led China to enter into agreements with unstable oil producers to maintain supplies at reasonable prices. The environmental impacts of China's energy consumption include air pollution and the increased contribution to global warming. Due to its reliance on coal for electricity, China has 16 of the 20 worst polluted cities in the world, causing its cities to become "invisible" due to the excessive smog (Fang 2006). The result has been increased attention by the government to the development of renewable and alternative energy sources. Furthermore, as the economic prospects of renewable energy development grow, the country, along with foreign investors and industries, is working to exploit the benefits they have to offer. As of 2006, China is the world's top investor in renewable energy, not including hydropower (Xinhua News Agency 2006). With this rate of investment, the Global Wind Energy Council predicts that China will surpass Germany, Spain, and the United States as the world's largest wind producer by 2020 (Xinhua News Agency 2007). The policies of the Chinese government are responsible for many of these developments.

On November 8, 2005, at the Beijing International Renewable Energy Conference, China pledged to the international community to double its reliance on renewable energy by 2020, which would bring renewable energy production up to 15 percent. To reach the 15 percent target, China has raised electricity rates to start a renewable energy development fund. It is up to the Chinese Renewable Energies Industry Association to continue developing a strategy under the Renewable Energy Law and to implement it. The Chinese Renewable Energy Industry Association was established in 2000 to provide institutional support for handling environmental problems caused by China's energy use through the development of the renewable energy sector.

Laws in China and other Asian countries are slightly different from Western laws. They are more like principles than actual plans for achieving the goals stated in the law. Chinese renewable energy experts say that, without an actual binding commitment to such targets, the goals are likely to go unmet or be reduced. But they are more likely to be met when they are expressed in a formal document like the Renewable Energy Law (Agence France Press 2006). In support of this, industry experts say that it is likely that China will reach its 2020 targets ahead of schedule (Worldwatch Institute 2006). It is notable, however, that the emphasis on renewable energy by the Chinese government is due more to the economic advantages associated with renewable energy than to the environmental benefits (Renewable Energy Access 2005).

The potential for renewable energy in China is great. According to Worldwatch, "an incomplete list of exploitable renewable energy reserves puts the country's hydropower potential at 400 million kW, wind at around 3 billion kW, biomass energy at 800–1,000 million tons of coal equivalent a year, and solar energy at a theoretical 1.7 trillion tons of coal equivalent a year" (Worldwatch Institute 2006). However, while renewable energies like wind, solar, biomass, and biogas are to be a part of the plan for meeting China's renewable energy targets, the major source of power will be hydroelectric. Development of large-scale hydroelectric resources has been a major priority of the Chinese government for over a decade. If fully exploited, hydropower in China could offset 40 percent of fossil fuel use ("Priority Given to Efficient Hydropower" 2004).

Chinese citizens have a different view of nuclear power than citizens in Western countries like the United States and many countries in the European Union. Unlike Western countries, where citizens have a very negative view of nuclear power, Chinese citizens are much more accepting. Even if the Chinese public did not have a high acceptance of nuclear power, it would be unlikely to have much of an effect on policy decisions. Decisions involving nuclear power in China are not open to public input (Florig and Zhao 2002, 61). The Chinese government is pursuing an aggressive strategy to increase nuclear power production, and, unlike many of the more developed countries in the world, public opinion is not a barrier. While nuclear is exalted by many supporters as an important solution to the problem of increasing carbon emissions, critics argue that, with China's growth rate, nuclear power will have a minimal impact on ghg emission reductions. Furthermore, the cost of disposal for nuclear waste could be problematic for future generations ("Oil Stock Boost for Emergency" 2007).

The future of China's energy situation hinges on many factors. First, it depends on whether the rate of economic growth continues at the expected pace. More or less growth will impact the energy needed, the funds available to develop more projects, and the ability of the government to regulate energy industries. Second, the level of international cooperation will have an impact on the amount and rate of energy projects that can be developed. China will need not only support for some of the technologies, but additional funding assistance in the form of aid or loans. Finally, renewable and alternative energy development hinges on the agenda of those in control of the government. The priority of hydropower development in the Chinese government has been attributed to the fact that Jiang Zemin, the former president, was an electrical engineer and Hu Jintao, the current president, studied hydroelectric engineering. The development of other types of renewable and alternative energy could also be dependent on the technical experience of members of the government. If governmental officials are experienced in other forms of renewable energy, there may be more support for its development. Furthermore, a shift in the country's paradigm that put more or less value on the economic and environmental outcomes of development could also affect the direction of the country's energy reliance. While many experts believe that China could become a leader in developing renewable and alternative energy, its reliance on fossil fuels will continue to increase as it copes with finding enough energy from any sources to provide electricity for its growth and the resulting improved standard of living (Miller 2006).

India

India, with the world's second-largest population and growth rate, has the potential to become the most populated country. Combine that possibility with its rate of economic growth, and India is expected to generate large increases in demand for energy over the next couple of decades (India Core 2006). This rate of economic and population growth makes it, like China, an important country for study in the field of energy production, use, and consumption. Despite the fact that it is not a fully developed country, India is the fifth largest consumer of energy in the world (Energy Information Administration 2007). The primary source of energy in India is coal, followed by oil and natural gas, though natural gas makes up only 7 percent of total energy consumed (Energy Information Administration 2007). Renewable energy, excluding hydropower, accounts for about 3.5 percent of total energy, hydropower accounts for almost 3 percent, and nuclear power accounts for about 2 percent, creating a great untapped potential for these energy sources (Energy Information Administration 2007).

Wind, solar, biomass, small hydro, and ocean energy have the potential to provide almost 30 percent, or the equivalent of 150,000 megawatts, of India's total energy needs (Meisen 2006,

20). Large-scale hydropower, if fully exploited, has the potential to provide over 10 percent, or the equivalent of about 57,000 megawatts, of total energy needs (World Bank 2007). And the potential of nuclear is, of course, limited to the amount of funding available for investment, uranium supplies, and those applications it is useful for. Because India is the only developing country that is producing hydrogen fuel cell vehicles, hydrogen has great potential there. The full exploitation of renewable and alternative energy could result in a major offset of fossil fuel use in India.

Energy has become a concern to the Indian government for many reasons. Like China, India has experienced fuel shortages and is even more reliant on energy imports. As a result, the president of India has made energy independence his top priority. To meet this priority, the Indian government will have to rely on a mixture of energy sources. It has low oil and natural gas supplies, but the third largest coal supply in the world. Although India is passing renewable and alternative energy policies, to what degree these sources will play into the energy mix is uncertain. Economic concerns tend to trump other concerns regarding the issue of energy. In addition to the energy independence priority, the Indian government has set a goal of 100 percent rural electrification by 2012. Renewable energy will be the cheapest and most efficient way to achieve this goal because many of these areas are located far from the energy grid (Energy Information Administration 2004a). Conclusively, renewable energy provides the benefit of a fully electrified nation, in particular, to India. As discussed in Chapter 2, using renewable energy to bring electricity to people in rural areas often results in higher equity throughout the world because people in poorer nations are able to enjoy the technological advances that more developed nations have been enjoying for decades.

It is evident that India's concern with renewable energy is not guided by its concern for the environment. As indicated by the statements of Surya P. Sethi, the principal energy policy adviser of India, the Indian government is more concerned with economic development ("Carbon Reduction or Poverty Reduction, Not Both" 2006). The goals of renewable energy policies in India are directed more toward the economic and standard-of-living advantages that using renewable energy provides, like achieving energy independence and bringing electricity to poor rural areas, than for the environmental benefits (Kalam 2006). This will have an effect on the rate and type of energy developed in India. Like China, India's laws are slightly different from Western laws. Targets for renewable energy development are more like suggestive targets and are not backed by any specific national policy. This means not that there are no policies in place for support, but that the goals are not necessarily binding and that other issues can overshadow efforts to develop renewable energy, including economics.

For years, India has had an aggressive renewable energy strategy and was one of the first developing countries to implement an electricity feed-in law. Most of the renewable energy policy was left up to the states with the passage of the Electricity Act of 2003, which also established a goal of an additional 10 percent of electricity from renewable energy nationwide. Despite the increased role of states in energy policy, the national government is still involved in the formulation of policies and targets. In 2006, Indian president Dr. A. P. J. Abdul Kalam called for an increase in renewable energy power generation from a previous target of 6 percent to 25 percent by 2030. Additionally, India is the first and only country to establish a Ministry of Renewable Energy, devoted solely to the development and advancement of renewable energy sources.

India has invested the most heavily in wind power. And it has recently surpassed Denmark to become the fourth largest producer of wind power in the world and now has the second fastest growing market for wind. India has one of the largest national programs to promote the development of solar power (Energy Information Administration 2004a). Small-scale hydropower is the most utilized source of renewable energy in India. Given that there is much potential to be tapped into, the Indian government is striving to develop its large-scale hydro projects. As the largest producer of sugarcane in the world, India has been implementing policies to take advantage of this considerable potential for the development of its biomass economy. Biogas or methane capture is also being pursued, with over 100,000 biogas plants installed in 2004–2005 alone (Renewable Energy Access 2005).

Like China, India views nuclear power as a viable source of energy for a country with a large and growing population. The nation has not developed its capacity for energy production like the more developed countries and also has limited funding sources to do so. As such, nuclear power is seen as a way to achieve goals of energy equity, rural electrification, and reliable supplies, which are seen as outweighing the costs of building the facilities, disposing of nuclear waste, and the risks of nuclear accidents.

Due to their similarities, India's use of renewable and alternative energy hinges on many of the same factors as China's. Regardless, India will be an important player in the advancement of renewable energy technologies. With the rate that these markets are growing in India, they will provide the market penetration that many renewable energy technologies need to bring the costs of the more expensive technologies down to a level that is more competitive with fossil fuel sources. The rural electrification program could serve as a model for other developing countries that are seeking to provide full electricity to all of their citizens. As already stated, concerns over the welfare of its people and economy are primary at this point in time. If India is to make environmental concerns related to energy a priority, it will be up to the developing nations to make this possible through financial, technical, and institutional assistance.

Japan

As one of the most highly developed countries in the world, Japan has the second highest rate of energy consumption and use in Asia and the third highest in the world (British Petroleum 2007). With its characteristic of being such a small but densely populated country, Japan relies heavily on imported energy because it does not have the fossil fuel resources to provide for its own energy needs. As a result, the oil shocks of the 1970s and 1980s prompted the government to diversify its energy supply and improve energy efficiency. While President Jimmy Carter's attempt to put the United States on the same path was thwarted by President Ronald Regan's policies, Japan followed through with its efforts and became the most efficient energy economy in the world (Moore and Ihle 1999). Through partnerships between the public and private sector, Japan emerged in the early 1990s as a leader in renewable energy and energy efficiency technologies alongside Germany. In 1997, Japan hosted the Kyoto session of the United Nations Framework Convention on Climate Change, where the Kyoto Protocol was developed as a binding amendment to the international treaty on climate change. Kyoto brought environmental concerns to the forefront, and Japan began implementing policies to encourage the market penetration of technologies that would reduce ghg emissions (Moore and Ihle 1999).

Oil and coal are the dominant sources of energy for Japan, making up about 47 percent and 21 percent of the total shares of energy, respectively. Natural gas accounts for only about 13 percent of the total energy supply, while nuclear power accounts for almost 14 percent and renewable energies, including large-scale hydropower, account for about 3.5 percent (European Commission 2006b, 2.8.2). If fully developed, it is estimated that renewable energies could greatly offset the use of fossil fuels in Japan's energy sector.

While nuclear power makes up a considerably larger share of energy in Japan compared to other developed countries, development of new nuclear power plants is still controversial. Published reports indicate that wind is a more viable fuel than plutonium for meeting Japan's energy needs (Fioravanti 1999). The Japanese government, however, views nuclear power as a necessary part of the fuel mix for achieving energy independence and security through a reduced reliance on fossil fuels. Government reports indicate that it is impossible for nuclear power to be phased out and replaced by renewable energy (Japanese Ministry of Economy, Trade and Industry 2006, 1).

Although policies to pursue the development of renewable energy technology began in earnest in Japan a couple of decades ago, policies to encourage the significant commercialization of renewable energy only emerged in 1998. The first notable policy to encourage the development of renewable energy technologies in Japan was the Sunshine Project, implemented in 1974 with a focus on solar PV technologies, geothermal, and hydrogen. In 1978, the Moonlight Project was implemented to encourage R&D of energy-efficient technologies. In 1980, the Law Concerning the Promotion of Development and Introduction of Petroleum Substituting Energy was passed, requiring that the government establish a workable plan that would include financial support for the actual use and deployment of renewable energy technology. This law also established the New Energy and Industrial Technology Development Organization (NEDO), a governmental organization that has been a key player in carrying out renewable and alternative energy policies. In 1992, the New Sunshine Law was passed, which combined the original Sunshine Project with the Moonlight Project and set the stage for policies that introduced net metering at competitive prices. Policies related to net metering coupled with ambitious national targets and the 70,000 Solar Roofs Program were the catalyst for the unprecedented growth of the PV market in the 1990s. Similar in many ways to the U.S. Million Solar Roofs Initiative, the Japanese Solar Roofs Program has seen more success because it set targets for energy production whereas the U.S. policy did not. Prior to 1997, the Solar Roofs Program was the only policy that encouraged the adoption of renewable energy technologies with a strategy that aimed to assist in market penetration. This policy was limited to PV solar technology, leaving other solar, wind, ocean, and biomass technologies at a disadvantage for market penetration.

In 1997, the New Energy Law was adopted. It created subsidies and other market incentives to encourage the adoption of these technologies by the public and private sectors. It is notable that the Ministry of International Trade and Industry and NEDO are the primary government agencies handling the implementation of Japan's renewable energy policies. What is even more interesting is that these agencies tend to employ a more collaborative relationship between the public and private sectors to achieve their policy goals. This kind of partnership can contribute greatly to the success of policies that deal with technology.

The Special Measures Law Concerning Promotion of the Use of New Energy by Electricity Utilities of 2002 was the first law to establish a market for green certificates in an effort to encourage the use of renewable energy through consumer choice. It also created a renewable energy portfolio that set a goal to increase renewable energy. The law has been criticized as setting too low a standard. Additionally, the law is not binding, making success uncertain at best (Iida 2002, 8).

Partly because of its ambitious and successful Solar Roofs Program, Japan dominates the PV market. Wind energy development has slowed due to a lack of incentives for investment in, development of, and adoption of wind energy. Due to the fact that Japan is a volcanic island country, it has significant geothermal resources. It is difficult to develop them, however, because many of the sites are located within national parks. It is likely that nuclear energy will continue to play an important role in the energy economy because there is more public acceptance for it in Japan. Japan is a leader in hydrogen fuel cell research and has invested heavily in R&D of hydrogen compared to other countries. Following in the same trend as other developed countries, Japan has recently shifted its policy focus from R&D to the encouragement of the commercialization of hydrogen fuel cells under the New Hydrogen Project (Hydrogen Association 2004, 2). The successful adoption of hydrogen as an energy carrier creates great possibilities for the role that renewable energy can play in its energy economy.

Australia

Although Australia is as advanced and developed as the United States or European Union, it uses less energy. Australia consumes about 5 percent of what the United States does (Energy Information Administration 2006, 303). This is because Australia has a lower population and a lower energy intensity. Australia, like the United States, is rich in all fossil fuel energy resources, and it has many of the energy problems associated with fossil fuels. Growing consumption and declining energy production have led to increased dependence on imported oil, raising concerns about the problems of relying on oil from unstable regions. Air pollution in the cities is worsening, bringing attention to the need for alternative fuels for transportation. Also, growing concerns about global warming are continuing to shape the agenda for renewable energy.

As a result, the government has taken a number of steps to reduce Australia's dependency on fossil fuels. Despite concerns about the environment, the economic issues surrounding energy are the driving force behind policy action. This is reflected in the programs that have been adopted. Australia did not ratify the Kyoto Protocol until 2008, after public officials responded to a shift in public attitude that was indicated by the 2007 election results. Any country that has ratified the Kyoto Protocol has exceeded the targets that are set by it. Because the increased use of renewable and alternative energy provides one of the more important solutions for meeting the ghg reduction targets of the protocol, countries that are not seeking to meet reduction targets are unlikely to have ambitious policies to increase renewable energy shares. At this point, Australia is no exception.

Australia's main source of energy is coal, which accounts for 45 percent of total energy consumption, followed by oil and natural gas at 33 percent and 19 percent, respectively. Hydropower provides 3 percent of the total energy, and nuclear and renewable energy sources furnish a negligible amount of energy, making Australia's renewable and alternative energy sector one of the most underdeveloped of the industrialized world (Energy Information Administration 2006, 329). Recent policies, however, reflect a move by the government to incorporate sustainable energy into the country's energy portfolio.

Most of the policies provide subsidies to encourage the private sector to invest in renewable energy. An exception to this is the Renewable Energy Act of 2000, which established the Mandatory Renewable Energy Target (MRET), requiring that about 5 percent of electricity be produced from renewable energy by 2010. A system of tradable renewable energy credits was established under MRET. The 5 percent target is criticized for being too low and nonbinding (Kent and Mercer, 2006, 1059). Also enacted in 2000 was the Renewable Remote Power Generation Programme (RRPGP) under the Measures for a Better Environment package. RRPGP is a rebate program that aims to subsidize the conversion of diesel-based generators to renewable energy generators, which those living in rural areas rely on and which are not connected to the electrical grid. A Photovoltaic Rebate Program took effect on January 1, 2000, and provides rebates for domestic and industrial installations of PV systems. The Renewable Energy Equity Fund (REEF) became effective in 1999 and provides funding assistance for R&D of renewable energy technologies. However, this program does not provide support for the market penetration of the technologies developed.

The 2004 white paper, "Securing Australia's Energy Future," is the most recent renewable energy policy action taken by the Australian national government. The white paper announced the enactment of several policies to encourage the development and commercialization of renewable energy. One of these initiatives is the Renewable Energy Development Initiative (REDI). REDI is a competitive, merit-based grant program that supports the early stages of commercialization in order to get technologies with potential the support they need to reach the market penetration stage. The Solar Cities Initiative is a series of trials in different cities that tests models for sustainable solar energy applications. The goal is to determine how to overcome barriers to the adoption of large-scale grid applications. Also under the white paper is the five-year, $20.4 million Advanced Electricity Storage Technologies Program. This program will identify and promote advanced storage technologies to increase the ability of renewable energy to contribute to the electricity supply. The Renewable Energy Industry Development, passed in 2003, is another measure that provides grants for projects that will address market barriers, surveys of renewable energy resources, and related topics. Ethanol production grants have been distributed since 2002 to maintain ethanol and biodiesel shares in the transport sector. While these policies are comprehensive and ambitious, they are not expected to make a significant impact on renewable energy shares in the market right away (Kent and Mercer 2006, 1060).

Australia has significant wind and solar resources, with limited large hydro resources. It also has geothermal resources. There are few policies to encourage the development of geothermal, because the sites are located too far from the grid. The potential capacity of PV technology far exceeds the electricity needs of the country, but issues of economic efficiency abound because this technology is also the most expensive. The Photovoltaic Rebate Program is essential for the development of this technology (Saddler, Disendorf, and Denniss 2004). Australia's stable political system, stable policies, relatively low levels of red tape, and flexible labor relations are key strengths in attracting investors of renewable energy, adding to the potential for success of these policies.

Other Notable Issues in Asia

Russia is unique in many ways as a developed country. With the fall of communism, Russia experienced a decline in energy consumption as the economy went into a recession. There are no legislative mandates like RPS or ARTs to support the use of renewable energy. There are, however, programs to encourage investment into R&D of renewable energy technologies. Russia has an extensive energy transmission system already in place, making it possible to bring energy produced from remote locations to more populated areas. Russia also has abundant renewable energy resources like wind, biomass, hydro, tidal, and even solar. Unfortunately, as long as there is economic instability, any major shifts in energy are unlikely (European Bank 2005).

Other developed countries in the region, like South Korea and New Zealand, have policies in place to promote renewable energy. Although neither has an RPS with mandatory targets, South Korea does have electricity feed-in laws in place. Both countries have some form of policy that either invests public funds directly into the renewable energy sector or encourages investment by the private sector through subsidies and tax breaks. Both North and South Korea also have considerable nuclear power developments, whereas New Zealand has none.

Many of the less developed Asian nations lack reliable or complete data on their energy production, use, and consumption patterns. This is also true for data on their renewable energy potential. It is known that biomass still provides a considerable amount of energy compared to what it provides for developed countries. Many developing Asian nations have not tapped into their hydropower potential and have prospects for large-, small-, micro-, and mini-scale developments. Without reliable assessments of the true potential of these resources to provide for the energy needs of these countries, little is being done in the way of investment or policy actions to encourage the development of renewable or alternative energy sources. Furthermore, many of the less developed countries are lacking experts with the technical knowledge necessary to maintain several of these complex systems. While many of these countries are not financially able to pursue programs that are as ambitious as those in China, India, Japan, and Australia, there is still a trend toward a renewable energy transition. There are several examples indicating that there is a future, albeit a slow moving one, for renewable and alternative energies.

Thailand is reliant mostly on oil, followed by natural gas, biomass, and then coal. Hydropower accounts for less than 1 percent of total energy (International Energy Association 2006). It is one of the most politically stable countries in the region, a necessary quality if a country is to have a successfully renewable and alternative energy plan. Compared to other nations, Thailand has some of the most aggressive policies in place to encourage renewable energy, including an RPS and ARTs to encourage the production of renewable energy for electricity use. It has only one nuclear reactor, which is for the purposes of research and provides a negligible amount of the total energy.

Cambodia has established a preliminary National Renewable Electricity Policy for the promotion of renewable energy as a source for rural electrification, much like India. The law, however, has not established binding goals or targets for the achievement of its goals. Its focus is on the fair distribution of electricity and the diversification of energy sources for the purposes of energy security rather than the environmental benefits of nonfossil fuel energy production. Many barriers must be overcome to achieve these objectives, the bulk of which will require the assistance of regional and global organizations as well as of individual countries. Assistance will need to be institutional, educational, scientific, and technical as well as financial (Australian Business Council for Sustainable Energy 2005).

Vietnam is still mostly reliant on combustible renewable energies, like biomass, as its primary energy source. Like India, it has an aggressive policy to bring electricity to about 5 million rural households. Recognizing the importance of renewable energy for electrifying rural households, this goal has become a driving force in the development of a renewable energy sector. The government has developed the Renewable Energy Action Plan (REAP) to help in the development of a renewable energy sector in Vietnam. The plan outlines the need for international assistance and a strategy of policy actions that need to be taken. The World Bank is working as a partner in REAP's 10-year plan along with many international actors. As the plan is still in the implementation stage, it is difficult to determine its success, though it is expected to result in the electrification of over 2.5 million households, or half of those in need of electricity (Australian Business Council for Sustainable Energy 2005, 10).

Africa and the Middle East

Africa is one of the more underdeveloped regions in the world. Many Africans live in poverty, lacking essentials such as clean water and electricity. Most countries in Africa are underdeveloped or developing with the exception of South Africa, which can be considered a newly industrialized nation on the brink of full development. Because of the lack of economic development, Africa tends to have a much lower energy demand and hence consumes much less energy than the rest of the world. Like some other underdeveloped regions in the world, many African countries rely on biomass for their basic energy needs like heating and cooking. Several of the utility companies are state owned, paying no taxes and generating many unpaid externalities. All of these characteristics contribute to the lack of attention by African governments to more complex renewable energy technology development. Most countries do not have policies in place to encourage renewable energy or energy efficiency, and they probably will not begin development of the renewable energy sector without international assistance. Even then, the economies of many countries will likely have to improve, and the basic needs of citizens will need to be met before renewable energy can become a major priority. Notably, because many countries rely on simple biomass resources for their energy needs, there has not been a considerable amount of investment in fossil fuel technologies, making it more feasible for the region to leapfrog these technologies and invest in more sustainable ones when it is ready to invest in energy technology.

Africa has an abundance of renewable energy resources with considerable potential for exploitation. Only about 7 percent of large-scale hydropower has been exploited. The mini and micro hydro potential has not yet been assessed. Of 9,000 megawatts of geothermal energy potential, only 60 megawatts, in Kenya, are being exploited (United Nations 2003, 1). The region has excellent solar energy resources compared to other areas, especially in the winter, giving it some of the greatest potential in the world for solar technology development. Many PV projects have been started throughout the continent to aid in rural electrification (Organisation for Economic Co-operation and Development 2004, 1). Due to Africa's location in relation to the tropical zones, it has much less wind potential than other regions. This has led to a standstill of the wind market in the region. Ocean energy has not been much a focus for the region, although in 1956 an ocean thermal energy conversion (OTEC) plant was under construction in the ocean near the Ivory Coast, but the project was halted because of a lack of funding (U.S. Department of Energy, Energy Efficiency and Renewable Energy 2006). There have been no significant investments in R&D of hydrogen-based technology, so technology transfer from more developed countries is essential for a speedy transition to a hydrogen economy in the future. If small-scale, local renewable energy is the primary focus of development, however, the need for hydrogen is less important because the energy produced for local use does not need to be transported.

A variety of issues have limited the success of renewable energy in Africa. Lack of economic development enhances the problems of the high costs of renewable energy, poor infrastructure, and the lack of skilled technicians and maintenance workers. Political instability contributes to the lack of government support, the lack of coordination and collaboration, and insufficient policy planning. For the development and dissemination of renewables to be successful, African countries need to formulate and implement long-term policies, determine the appropriate types, develop capacity-building programs specific to those technologies, and tap into internationally supported financing opportunities.

South Africa

South Africa is one of the most developed countries in Africa. Its energy usage is more similar to industrialized nations than to its underdeveloped counterparts on the African continent. For instance, South Africa consumes 45 percent of the total energy in Africa (United Nations 2003, 3). Although it still relies heavily on biomass as a source of energy, South Africa also relies on coal for 90 percent of its electricity production and has developed nuclear power. Only 0.01 percent of its large-scale hydropower is being exploited. It has the highest wind potential of all the countries in Africa, although little of this has been developed (United Nations 2003, 15). Recognizing the impacts of its energy use and the benefits of sustainable energy development, the South African government has developed a more enthusiastic energy plan that aims at promoting a transition to renewable and alternative energies.

In 2003, the government published the "White Paper on Renewable Energy," which addresses the need for policy support of renewable energy. The white paper establishes a low target for renewable energy production by the year 2013 and focuses on the development of a variety of renewable energy technologies. The policy seeks to address many of the barriers to renewable energy market penetration and focuses on the need for research and the development of renewable energy. Rural electrification is a policy goal for South Africa and has resulted in the highest growth in the PV market of any African country. Despite this growth, PVs have not been found to be the most economically efficient solution for rural electrification. This has caused nuclear power development to gain status on the political agenda. Due to its status as a newly industrialized nation, economic efficiency will continue to play a role in the energy plans of the future (Energy Information Administration 2004b).

Middle East

The Middle East is well-known for its abundant oil reserves. These reserves are the driving force behind energy technology development in the region. Although the area is rich with this resource, growing demand is starting to catch up with the supply. This is leading to an increased concern over the development of sustainable energy in the region. From the development of re-

newable energy research centers to initiatives like the Trans-
Mediterranean Renewable Energy Cooperation, actions are being
taken by domestic and foreign actors to help with the development of renewable and alternative energy sources.

The Middle East has large solar resources and opportunities
for wind power (International Energy Association 2006). As such,
there is a considerable amount of investment in solar energy. The
United Arab Emirates is investing heavily in large- and small-
scale solar power as well as in wind farms. The organization recently built the first wind power project of the region on Sir
Baniyas Island and has plans for the construction of more wind
farms with technology built specifically to withstand the heat
and humidity of the area. Although it has no major investments
yet, Saudi Arabia is looking into developing its renewable energy
resources. Iran has plans to exploit its wind, solar, and geothermal resources. The development of nuclear power in the region
has caused international tensions, especially in the case of Iran.
Israel is the only country to pass an RPS, albeit for only 5 percent
of electricity shares by 2016. Other countries in the area have
begun doing studies to determine the potential of renewable energy resources. Although the region does not have the most aggressive policies in the world, its financial resources give it the
ability to make the investments needed for faster development.
The political instability of the region and its reliance on oil may
prove to be the greatest barrier to the realization of its renewable
and alternative energy capacity.

North America

North America has promising potential for renewable and alternative energy development, as Chapter 2 demonstrated about the
United States. It has abundant renewable energy supplies and
meets many of the other conditions necessary for full utilization
of the resources. Between Canada and the United States, the technological know-how for both renewable resources, nuclear and
hydrogen, are among the best in the world. Cooperative frameworks for Canada, the United States, and Mexico have already
been established through trade regimes like the North American
Free Trade Agreement and other regional agreements. The potential for this region to overcome the barriers to renewable energy
development discussed in Chapter 2 is still uncertain. The United

States remains the highest consumer of energy in the world, but has relatively low federal involvement in the encouragement of market penetration for renewable energy technologies. As a developing country, Mexico needs more international assistance than the United States and Canada to fund its renewable energy projects. The region of Central America remains relatively undeveloped compared to the United States, Canada, and Mexico and is definitely reliant on international technology sharing and financing to exploit its renewable energy resources.

Canada

As one of the more developed countries in the world, Canada ranks eighth in terms of energy consumption (International Energy Association 2006, 50). It ranks fifth in the world in terms of carbon dioxide emissions (International Energy Association 2006, 50). Unlike the United States, Canada has committed to the targets set by the Kyoto Protocol. Not having the strong support of state governments for the protocol as in the United States, the Canadian federal government faced opposition from its provinces when it signed the agreement. Since 2001, a few provinces have begun to pass policies to meet the targets. Canada will be using renewable and alternative energy sources as a way to reduce carbon dioxide emissions to the required targets under the agreement. With concerns growing over the environment, energy security, and the economy, the Canadian government and several provincial governments have implemented policies to encourage the development and market penetration of renewable and alternative energies.

Canada, like many other developed nations, relies primarily on oil, which accounts for over 36 percent of the total energy the country consumes (International Energy Association 2006). Compared to many other developed nations, Canada uses much less coal, about 10 percent, and its renewable energy shares are higher than average. Renewable energy provides just over 15 percent of the total energy used (International Energy Association 2006). In terms of production, Canada is the fourth-ranking producer of large-scale hydroelectric power, behind China (REN21 2006, 3). In fact, over 55 percent of its electricity needs are met with large-scale hydropower (International Energy Association 2006, 19). Wind, solar, ocean, and geothermal are extremely small contributors to the total energy capacity. Ethanol is playing an increasing role with new plants being built and federal laws that make ethanol account for 10 percent of the motor fuel mixture. Nuclear power makes up almost 9 percent of the total energy.

As discussed in Chapter 1, Canada is well-known for its nuclear technology development and exports its breeder reactor knowledge to many other countries. While there is controversy about using nuclear power in Canada, its citizens have been much more accepting. Whereas the United States stopped licensing new nuclear plants, Canada has continued bringing plants online throughout the 1990s and into the new millennium. The upkeep of aging plants and the construction of new plants will continue indefinitely.

Canada is very similar to the United States in terms of its policy actions toward renewable and alternative energy. Like the United States, the Canadian federal government has not established a national RPS, but three Canadian provinces have. The federal government did, however, pass the Wind Power Production Initiative (WPPI) in 2002, which is a national electricity feed-in law specific to wind power. It has a goal of increasing wind power shares in electricity by 500 percent over a few years, but it does not set mandatory targets for the public or private sector. Two provinces have implemented new electricity feed-in laws for a variety of renewable energies since 2004 (REN21 2006, 9). There is also a trend among the provinces to impose renewable fuel requirements for the transportation sector. Most federal policy focuses on providing incentives for the private sector to develop and utilize renewable energy technologies. Canada is one of the few countries in the world with serious hydrogen R&D funding. In collaboration with the private sector, the Canadian government has set aside millions of dollars to fund its hydrogen projects (Solomon and Banerjee 2006, 89). There is also a focus in many of the national policies on increasing public awareness of renewable and alternative energy sources.

Despite the progress that Canada has made in establishing policies to support renewable energy, it is behind many other developing countries in using these technologies to their potential. Little is being done to ensure the market penetration of technologies like wind, solar, and ocean. As discussed throughout the book, government intervention is necessary for the full penetration of renewable energy into the market. By placing their investments in nuclear power, private sector utilities are diverting funds from renewable energy technologies. Like so many other countries, Canada's renewable energy market could benefit from the incorporation of its positive externalities and an accounting of the costs and negative effects of energy sources like oil, coal, large-scale hydro, and nuclear. It is unlikely that the growth of renewable energy will improve without the government creating more targets and mandates.

Mexico

Mexico falls into the same category as many other newly industrialized nations. It uses much less energy than more developed nations, but its economic growth will cause its energy consumption to increase, creating a concern for the future of its energy supply. Less concerned with the environment than it is with economic efficiency, it falls victim to the classic dilemma associated with renewable and alternative energy: a lack of funding due to the high initial short-term costs. Mexico remains reliant on the advances of other countries in the development of renewable energy technology. It requires information sharing about technologies because it has little to contribute to R&D programs itself. Regardless, the Mexican government is still passing policies to encourage renewable energy where it can.

The Mexican economy is dependent on oil for almost 60 percent of its energy needs. This percentage is much higher than most countries in the world (International Energy Association 2006). Natural gas ranks second at over 26 percent. Unlike most countries, renewable energy sources provide more of the total energy than coal (almost 10 percent), which accounts for less than 5 percent of total energy (International Energy Association 2006). While the majority of Mexico's renewable energy shares are produced by large-scale hydro, its geothermal resources are also fairly developed. Mexico ranks third in the world for geothermal energy production (REN21 2006, 3). But like most countries in the world, its other renewable energy sources are significantly undeveloped. Nuclear power provides less than 2 percent of Mexico's energy needs. These facts, combined with the problems surrounding the reliance on imported oil, have brought the potential of renewable energy as a solution for improved energy security to the forefront. Policies that encourage renewable energy as a solution have been passed over the last decade.

Renewable energy policy is at an early stage in Mexico. Laws to encourage the private investment in the use of renewable energy technologies have been in place the longest. As recently as 2003, laws allowing net metering for renewable energy production were passed. Electricity feed-in laws and RPSs have not been incorporated into the policy mix, but policies have been passed requiring that these requirements be developed. It is uncertain as to when binding targets will be passed in the future or how ambitious they will be. There are no plans for the development of nuclear power. The one plant in operation is not viewed by the government as having benefits that outweigh the costs of a project in operation for more than 30 years. Furthermore, funding is limited for development.

The future of renewable energy in Mexico hinges on overcoming several of the same barriers that many developing and developed countries must overcome. Mexico is on the right track, though a slow one, in creating the support needed for the success of renewable energy penetration into the market. It will, however, be dependent on international financing and technological assistance until it is financially stable enough to fund R&D on a large scale. The political stability of the country creates an environment with more potential for success. While Mexico will not be a leader in renewable energy production anytime in the near future, it will be receptive to the leadership of other countries to a transition to a more sustainable energy economy.

Central America

Central America consists of Belize, Costa Rica, El Salvador, Guatemala, Honduras, Nicaragua, and Panama, which are all developing countries, though the region is considered one of the poorest in the world. While it does have many natural resources, it has relatively few fossil fuel energy resources. This emphasizes the need for the development of renewable energy in the region, especially to aid in the electrification of rural areas. Small efforts by a few countries are being made to increase the renewable energy shares, but their success hinges on the amount of support they receive from the international community.

Oil is the primary energy source for most countries, making up 70 percent of the total, and pollution from cars and industry is rapidly becoming a problem in some areas. Large-scale hydropower contributes the second highest amount of energy to the region, 22 percent (Energy Information Administration 2002). Although hydropower is an important resource for the region, the weather patterns create risks for excessive rain and flooding, which has resulted in evacuations that would be affected by the breaching of a dam. The first coal-powered plant was not built until 2000 in Guatemala and therefore provides hardly any of the power for the region. At the time of publication, energy from natural gas and nuclear sources was not used in this region at all. The development of complex renewable energy technologies like solar, wind, and biomass has only begun over the last decade and only with international assistance.

Most renewable energy development is driven by rural electrification programs seeking to provide electricity for the entire population, similar to many developing countries in Asia. Honduras has a notable rural electrification program with plans to build renewable capacities to reach its goals. Costa Rica has the most advanced renewable energy economy with geothermal and biomass developments accounting for almost 50 percent of total energy use. Finland has established a partnership with the governments of Central America to help with the development of renewable energy. The state of economic development in the region, or the lack thereof, will certainly prohibit serious development of renewable and alternative energy in the region unless international assistance is provided for that aim.

South America

South America has a promising renewable energy market, but it also has some of the world's largest oil reserves. The combination has the potential to be both helpful and harmful to the development of renewable sources. On the one hand, the existence of oil reserves can cause a natural dependence on those resources while creating a political environment full of conflict between those who want to use the oil to meet the economic needs of the region and those who recognize the need for a transition to more sustainable energy sources. Furthermore, a strategy of exporting oil while reducing dependence on it domestically has the potential to generate the money needed for the development of a renewable energy market. This is a major issue for the region because Venezuela, which has large oil resources, has been attempting to undermine the development of Brazil's ethanol market. The political situation between Brazil and Venezuela is complex. Both countries are exporting a competing product, and Venezuela has criticized Brazilian plans for increased ethanol production, arguing that it could create risks for food security. The conflict, however, involves disagreements not just over which energy source to use, but also over exportations of both fuels to foreign countries. Venezuela does not want Brazil to be involved in a cooperative effort with the United States to provide ethanol for U.S. energy needs because of the tense relations between the United States and Venezuela. Brazil would like to export ethanol in an effort to boost its economic development. At the First South American Energy Summit on April 16, 2007, Venezuela discussed the possibility of importing ethanol from Brazil so that it does not have to rely on the U.S. market for exports (Sanchez 2007). Despite Venezuela's opposition to Brazil's plans for ethanol production, it does not appear that Brazil will change its energy policy.

Brazil

Despite its status as a developing country, Brazil has promoted renewable energy development in many ways that even developed countries have not. The share of renewables in its energy market are slightly less than that of oil and much higher than those of coal and natural gas. This is mainly due to its intensive development of biofuels that began in the 1970s after the first oil shocks. It now has one of the most promising renewable energy markets in the world. In fact, Brazil is the highest-ranking producer of ethanol on earth, sharing the ranking with the United States after years as the top producer. There has been a lack of development of other renewable energy sources because all of the funding available for energy development has been directed to large-scale hydro projects and ethanol production. Recent policies show a shift in priorities in Brazil toward the development of a more diverse energy economy. Brazil is expected to rank with China and India as one of the leading renewable energy markets of the future.

Receiving much of its energy supply from renewable resources, Brazil relies on hydropower for over 80 percent of its electricity needs (International Energy Association 2006, 19). Therefore, it ranks third in the world for electricity produced from large-scale hydro sources, and it is also one of the highest-ranking small-hydro energy producers (REN21 2006, 3). Ethanol provides over 40 percent of its motor vehicle fuel consumption (REN21 2006, 5). The use of flex-fuel vehicles, which can run on both ethanol and gasoline, makes this possible in Brazil, where such vehicles make up over 70 percent of the vehicle market (REN21 2006, 5). Oil accounts for just over 43 percent of the total energy supply, and renewable energy accounts for 40 percent. Coal and natural gas account for over 7 percent each of the energy supply, and nuclear energy sources account for less than 2 percent. Although nuclear power does not contribute much to the energy mix, Brazil has some of the world's largest uranium deposits. Solar and wind power have not been developed on a large scale anywhere in the country because of the focus on biofuels and hydropower development. Hydrogen research is in its early stages in this country, but, with its biomass resources, Brazil is one of two countries in the world with the potential to produce hydrogen using only renewable energy (Solomon and Banerjee 2006). The political actions and climate in the near future will determine whether Brazil reaches its potential for renewable energy production.

The National Alcohol Program of 1975 was implemented after the oil crisis of 1973–1974 and led to the largest-scale development of alcohol-based fuels, including ethanol, in the world (Solomon and Banerjee 2006). Recent federal mandates show a continuation of this trend by setting a target to increase sugarcane production for ethanol by 40 percent by 2009. In addition, Brazil is spending more than any other country on its 2008 plans for the construction of new biofuel production plants.

In 2006, Brazil announced plans to build seven new nuclear power plants by 2025. This would be in addition to the two plants already producing power and a third that is expected to be finished in 2010. As a result, nuclear power would increase its share of the electricity market by 3 percent. In the past, issues with funding have delayed nuclear development, and it is possible that the projects could be delayed again. The launch of a new uranium enrichment center in 2006 could aid in Brazil's nuclear ambitions because the nation will no longer have to rely on other countries for enriched uranium. Concerns have arisen in the international arena over the political consequences of these actions, given the tense relationship between Venezuela and Brazil. As Brazil moves closer to energy independence through the development of nuclear resources, there may be a move by Venezuela to develop nuclear power. In the end, both countries could be capable of building a nuclear weapons arsenal, which could greatly strain the increasingly unstable relations between the two.

In the case of other renewable energy sources, the PRO-FINA program provides a set of stages for various policy mechanisms with an aim to increase the production of energy from renewable energy sources like wind, bioenergy, and hydropower. In English, PROFINA stands for the Program of Incentives for Alternative Electricity Sources. The policy mechanisms include electricity feed-in laws, RPS obligations, subsidies and tax breaks, and tradable permits. This policy encourages the development only of certain renewable energy technologies, neglecting to provide incentives for solar and ocean energy development. Such sources may be developed under Brazil's rural electrification programs, though it is doubtful that these more expensive technologies will be invested in to the extent that the other renewable energy technologies have.

The result of these policies will undoubtedly lead to an increase in the use of renewable and alternative energy. The lack of attention to the broad array of technologies available could result in an unbalanced reliance on some sources, like large-scale hydro, ethanol, and nuclear power. While this could certainly result in more positive environmental effects than the use of fossil fuels, all these sources have the greatest potential for environmental and human harm of all the renewable and alternative energy technologies. The political instability of the region could also be worsened by the growing energy independence of Brazil. Because it is almost certain that Brazil will play a great role in the future of energy markets, international assistance to help with peacekeeping efforts could be more important than economic and technological assistance.

Other Issues in South America

Most countries in South America have not passed any significant policies to develop renewable energy, although Paraguay and Argentina are exceptions to the rule.

Paraguay, in an effort to increase energy security and rural electrification, has implemented policies aimed at diversifying the energy mix by working to develop the regulations required for the proper functioning of the energy market when production from new sources is added. While renewable energy accounts for over 80 percent of the total primary energy supply, the majority of this is from biomass. The negative effects on the environment from using primarily biomass to supply energy for a more developed economy has also caused the Paraguayan government to shift its focus to more sustainable energy sources like wind and solar power.

Argentina's rural electrification target prompted the passage of a renewable energy target of 8 percent. As of 2006, the policy has resulted in an increased capacity for renewable energy production and is expected to continue contributing to the growth of renewable energy. Bolivia has a rural electrification program as well, which has a target of 50 percent access by 2015 and full access by 2025. Bolivia has plans to build 20,000 solar home systems as part of the program by mid-2009—one of the most ambitious solar energy programs in the region. The rural electrification plans in Peru and Chile also recognize the potential of renewable energy for achieving its goals. In these countries, the passage of regulations will be essential for successful penetration of renewable energies into the mainstream energy market. Of course, international assistance in the form of funding, technology sharing, and policy advice will also increase the chances of success.

Conclusion

Given the high initial expense of renewable energy projects, even developed countries are struggling to get the investment funds needed by the private and public sectors to make a speedy and successful transition to a more sustainable energy economy. As discussed in Chapter 2, this is a problem in the developed world because of the dominance of the fossil fuel industry in the economy and because of the misperceptions about the true costs of using renewable and alternative energy compared to traditional sources. This translates into a problem for developing and underdeveloped countries that almost certainly require financial assistance from the developed world to enable renewable energy resources to grow. Although developing countries require financial assistance from the international community, many of these countries have the advantage of lacking an energy structure to begin with. As these countries seek to provide electricity for populations that have never had it, they can leapfrog over inefficient technologies and technologies relying on traditional fossil fuel energy sources. Although the initial investment may be high, the development of renewable energy in the Third World has many benefits, including energy security, political stability, and environmental sustainability.

To begin to achieve a renewable energy transition throughout the world, policymakers must make changes that decrease the unfair economic power of fossil fuel industries and that reflect the true costs and benefits of all energy sources. Additionally, due to the global nature of many of the negative effects of traditional energy technologies, it will be crucial for more advanced countries to provide assistance in all forms to less advanced countries to help prevent further damage to the environment. If these actions are not taken, the prospects become bleak that a transition to a renewable energy economy will occur within the time frame needed to prevent the destruction caused by fossil fuel use, as described in Chapter 2. The next chapter provides a chronology of the events that have contributed to our understanding of the energy situation in this context, as well as the technological breakthroughs that have made renewable and alternative energy sources an option.

References

Agence France Press. "China Lowers Target for Renewable Energy." October 26, 2006. Lexis-Nexis. www.lexisnexis.com

Ames, Paul. "EU Drafts Plan to Battle Global Warming; Sets Binding Targets for Switch to Green Energy." *Environmental News Network*. March 9, 2007. www.enn.com/energy/article/6160

Asia Pacific Partnership on Clean Development & Climate. "Renewable Energy and Distributed Generation Task Force Action Plan." November 2007. www.asiapacificpartnership.org/RenewableEnergyTF.htm

Australian Business Council for Sustainable Energy (BCSE). "Renewable Energy in Asia: The Cambodia Report." Carlton, Victoria: Author, August 2005a. www.bcse.org.au/docs/International/BCSE%20Cambodia %20Final%20V2.pdf

Australian Business Council for Sustainable Energy (BCSE). "Renewable Energy in Asia: The Vietnam Report." Carlton, Victoria: Author, August 2005b. www.bcse.org.au/docs/International/BCSE%20Vietnam%20%20 Final%20V2.pdf

"Authorized Release: The Renewable Energy Law." Renewable Energy Access. No date. www.renewableenergyaccess.com/assets/download /China_RE_Law_05.doc

British Petroleum. "Statistical Review of World Energy 2006." 2007. www.bp.com/statisticalreview

"Carbon Reduction or Poverty Reduction, Not Both." Reasononline. November 14, 2006. www.reason.com/news/show/116724.html

Chang, J., Dennis Y. C. Leung, C. Z. Wu, and Z. H. Yuan. "A Review on the Energy Production, Consumption and Prospect of Renewable Energy in China." *Renewable and Sustainable Energy Reviews* 7, 5 (2003): 453–468.

"China's Hydropower Resources Top World." *China Daily.* May 30, 2004. search.chinadaily.com.cn/searchen.jsp?channelname=ICS_EN&strTodat e=2004-05-30&secondsearch=ON&strFromdate=2004-05-30&search-Text=hydropower&ch=all&usedate=yes&f_yy=2004&f_mm=05&f_dd=3 0&t_yy=2004&t_mm=05&t_dd=30&strAuthor=&Submit=Search

Clean Development Mechanism in China. "China Accelerates Construction of Renewable Energy Projects." August 1, 2006. cdm.ccchina.gov.cn /english/NewsInfo.asp?NewsId=1068

Clean Development Mechanism in China. "Approval Status of CDM Projects in China." May 10, 2007. cdm.ccchina.gov.cn/WebSite/CDM /UpFile/File1277.pdf

Commission of the European Communities. "Communication from the Commission to the Council and the European Parliament. Renewable Energy Road Map: Renewable Energies in the 21st Century: Building a More Sustainable Future." Report 848. January 10, 2007a. ec.europa.eu /energy/energy_policy/doc/03_renewable_energy_roadmap_en.pdf

Commission of the European Communities. "Communication from the Commission to the Council and the European Parliament. Green Paper Follow-up Action: Report on Progress in Renewable Electricity." Report 849, January 10, 2007b. 72.14.253.104/search?q=cache:hbyRFTMbNeQJ :ec.europa.eu/energy/energy_policy/doc/06_progress_report_renewable _electricity_en.pdf+EU+renewables+directive+progress&hl=en&ct=clnk &cd=2&gl=us&client=firefox-a

Cook, Lorne. "EU Split on Renewable Energy Targets." *Energy Daily.* March 5, 2007. www.energy-daily.com/reports/EU_Split_On_Renewable _Energy_Targets_999.html

Danish Energy Authority. "Energy Strategy 2025." 2005. www.ens.dk /graphics/Publikationer/Energipolitik_UK/Energy_Strategy_2025/ index.htm

Department of Trade and Industry (DTI). "UK Energy in Brief, July 2006." National Statistics Publication. 2006. www.berr.gov.uk/files /file32387.pdf

Ecoworld. "India's Nuclear Power, Assisting Energy Independence or a Dangerous Experiment?" 2005. www.ecoworld.com/home/articles2.cfm?tid=402

Energy Information Administration (EIA). "Executive Summary." *Impacts of the Kyoto Protocol on U.S. Energy Markets and Economic Activity.* October 1998: xi. tonto.eia.doe.gov/ftproot/service/oiaf9803.pdf

Energy Information Administration (EIA) "Central America Regional Indicators." August 2002. www.eia.doe.gov/emeu/cabs/guatemal.html

Energy Information Administration (EIA). "Country Analysis Briefs: India." 2004a. www.eia.doe.gov/emeu/cabs/India/pdf.pdf

Energy Information Administration (EIA). "Country Analysis Briefs: South Africa." 2004b. www.eia.doe.gov/emeu/cabs/South_Africa/Full.html

Energy Information Administration (EIA). *Annual Energy Review 2005.* 2006. tonto.eia.doe.gov/FTPROOT/multifuel/038405.pdf

Energy Information Administration (EIA). "Country Analysis Briefs: India." 2007. www.eia.doe.gov/emeu/cabs/India/Background.html

Euractiv. "EU Divided over Nuclear and Renewables." March 8, 2007a. www.euractiv.com/en/energy/eu-divided-nuclear-renewables/article-162309

Euractiv. "EU Makes Bold Climate and Renewables Commitment." March 29, 2007b. www.euractiv.com/en/energy/eu-bold-climate-renewables-commitment/article–162373

European Bank for Reconstruction and Development. "Russia: Country Profile 2005." 2005. www.ebrdrenewables.com/sites/renew/countries/Russia/profile.aspx#Policy

European Commission. "State of Renewable Energies in Europe—2006." 2006a. www.energies-renouvelables.org/observ-er/stat_baro/barobilan/barobilan6.pdf

European Commission. "Part 2—Energy." In *Energy and Transport Figures in 2006.* Luxembourg: European Commission/Eurostat, 2006b. bookshop.europa.eu/eGetRecords

European Commission. "France—Renewable Energy Fact Sheet." January 2007a. ec.europa.eu/energy/energy_policy/doc/factsheets/renewables/renewables_fr_en.pdf

European Commission. "Netherlands—Renewable Energy Fact Sheet." January 2007b. ec.europa.eu/energy/energy_policy/doc/factsheets/renewables/renewables_nl_en.pdf

Fang, Bay. "China's Renewal." *U.S. News & World Report* (January 12, 2006). Lexis-Nexis. www.lexisnexis.com

Federal Ministry of Economics and Technology. "Renewable Energies in Germany—A Success Story." August 7, 2007. www.german-renewable-energy.com/Renewables/Navigation/Englisch/root.html.

Florig, H. K., and S. Zhao. "Debating Perspectives on Nuclear Power Safety." *Science and Technology Review* 165 (March 2002): 61–64.

Fioravanti, Marc. "Wind Power versus Plutonium: An Examination of Wind Energy Potential and a Comparison of Offshore Wind Energy to Plutonium Use in Japan" (report prepared for the Institute for Energy and Environmental Research). January 1999. www.ieer.org/reports/wind/marcstat.html

Guardian. "China Pledges to Double Reliance on Renewable Energy by 2020." November 8, 2005. www.guardian.co.uk/china/story/0,7369,163 6632,00.html

Government Offices of Sweden. "Energy Policy." 2007. www.sweden.gov.se/sb/d/5745/a/19594

Holloway, T., A. Fiore, and M. Galanter-Hastings. "Intercontinental Transport of Air Pollution: Will Emerging Science Lead to a New Hemispheric Treaty?" *Environment Science and Technology* 37, 20 (2003): 4535–4542

Hydrogen Association. "A Global Hydrogen Effort—Factsheet." August 2004. www.hydrogenassociation.org/general/factSheet_global.pdf

Iida, Tetsunari. "Green Power and Renewable Energy Policy in Japan." (Presented at the 7th National Green Power Conference.) September 30, 2002. www.eere.energy.gov/greenpower/conference/7gpmc02/iida02.pdf

India Core. "Energy Overview." 2006. www.indiacore.com/overview-energy.html

International Energy Association (IEA). "World Energy Outlook 2006." 2006. www.worldenergyoutlook.org

International Energy Association (IEA). "Global Renewable Energy Policies and Measures Database." 2007. www.iea.org/textbase/pm/grindex.aspx

International Panel on Climate Change (IPCC). "Summary for Policy Makers." *Climate Change 2007: The Physical Science Basis.* February 2007. www.ipcc.ch/pdf/assessment-report/ar4/wg1/ar4-wg1-spm.pdf

Japanese External Trade Organization (JETRO). "Wind Power on the Increase in Japan." *Japan Economic Report.* April–May 2006. www.jetro.go.jp/en/market/report/pdf/2006_17_hs.pdf

Japanese Ministry of Economy, Trade and Industry. "Main Points and Policy Package in 'Japan's Nuclear Energy Plan'" (report from the Nuclear Energy Subcommittee). September 2006. www.enecho.meti.go .jp/english/report/rikkokugaiyou.pdf

Kalam, A. P. J. Abdul. "Address at the Inauguration of the South Asian Conference on Renewable Energy New Delhi." April 18, 2006. mnes .nic.in/president/splangnewPDF%2520Format758.pdf

Kent, Anthony, and David Mercer. 2006. "Australia's Mandatory Renewable Energy Target (MRET): An Assessment." *Energy Policy* 34, 9 (2006): 1046–1062.

MacGill, I., M. Watt, and R. Passey. "The Economic Development Potential and Job Creation Potential of Renewable Energy: Australian Case Studies." Report Commissioned by Australian Cooperative Research Centre for Renewable Energy Policy Group, Australian Ecogeneration Association and Renewable Energy Generators Association. 2002.

Meisen, Peter. "Overview of Renewable Energy Potential of India." October 2006. Global Energy Network Institute (GENI). www.geni.org /globalenergy/library/energytrends/currentusage/renewable /Renewable-Energy-Potential-for-India.pdf

Miller, Tom. 2006. "Greener China with Double the Coal?" *FT.com* (November 5, 2006). us.ft.com/ftgateway/superpage.ft?news_id=fto110520 062339123456

Moore, Curtis, and Jack Ihle. "Renewable Energy Policy Outside the United States." Renewable Energy Policy Project. 1999. www.crest.org /repp_pubs/articles/issuebr14/01Intro.htm

National Audit Office. "Executive Summary." In *Renewable Energy*. Department of Trade and Industry, Report by the Comptroller and Auditor General. February 2005. www.nao.org.uk/publications/nao_reports/04-05/0405210es.pdf

"Oil Stock Boost for Emergency." *China Daily*. April 23, 2007. www1.cei .gov.cn/ce/doc/cenm/200704231627.htm

Organisation for Economic Co-operation and Development (OECD). "Exploiting Africa's Huge Energy Potential as a Weapon Against Poverty." 2004. www.oecd.org/dataoecd/43/45/32285615.PDF

"Priority Given to Efficient Hydropower." *China Daily*. October 31, 2004. www.chinadaily.com.cn/english/doc/2004-10/31/content_387070.htm

Renewable Energy Access. "Snapshot of Renewable Energy in India." May 23, 2005. www.renewableenergyaccess.com/rea/news/story?id =30729

Renewable Energy Policy Network for the 21st Century (REN21). *Renewables Global Status Report 2006 Update*. 18 July 2006. www.ren21.net /globalstatusreport/g2006.asp.

Runci, Paul. "Renewable Energy Policy in Germany: An Overview and Assessment." Pacific Northwest National Laboratory Technical Report PNWD-3526. January 17, 2005. www.globalchange.umd.edu/energy-trends/germany/1/

Saddler, Hugh, Mark Disendorf, and Richard Denniss. "A Clean Energy Future for Australia." March 2004. wwf.org.au/ourwork/climatechange /cleanenergyfuture/

Sanchez, Fabiola. "Chavez Seeks to Counter U.S. Influence at South American Energy Summit." *Indian Country Today*, April 23, 2007. www.indiancountry.com/content.cfm?id=1096414885

Sijm, J. P. M. "The Performance of Feed-in Tariffs to Promote Renewable Electricity in European Countries." Report ECN-C-02-083. 2002. Petten, The Netherlands: Netherlands Energy Research Foundation, November 2002.

Solomon, Barry D., and Abhijit Banerjee. "A Global Survey of Hydrogen Energy Research, Development and Policy." *Energy Policy* 34, 11 (11 July 2006) 781–792.

United Nations. "Renewable Energy in Africa: Prospects and Limitations." 2003. www.un.org/esa/sustdev/sdissues/energy/op/nepad-karekezi.

U.S. Department of Energy, Energy Efficiency and Renewable Energy. "Ocean Thermal Energy Conversion." 2006. www.eere.energy.gov /consumer/renewable_energy/ocean/index.cfm/mytopic=50010

World Bank. "Hydropower Development in India." 2007. www.world-bank.org.in/WBSITE/EXTERNAL/COUNTRIES/SOUTHASIAEXT /INDIAEXTN/0,,contentMDK:21388713~pagePK:141137~piPK:141127 ~theSitePK:295584,00.html

Worldwatch Institute. "China Speeds up Renewable Energy Development." October 26, 2006. www.worldwatch.org/node/4691

Xinhua News Agency. "China a Leading Investor in Renewable Energy." May 17, 2006. www.china.org.cn/english/environment/168626.htm

Xinhua News Agency. "China to Become World's Largest Wind Power Producer." January 7, 2007. www.uofaweb.ualberta.ca/chinainstitute /nav03.cfm?nav03=55229&nav02=50883&nav01=43092

4

Chronology

Many events have led up to our current energy situation. Technological breakthroughs and discoveries seem to be the main force driving the insatiable demand for energy and the subsequent transitions to different energy sources. More recently, discoveries about the environment, including the effect of greenhouse gases on the global climate, have played a key role in the development of renewable energy. This chronology provides a comprehensive list of the events that have had an impact on world energy issues. In particular, the events are related to renewable and alternative energy. The use of energy allowed humans to bring about the sophisticated and complex civilization of today. Ironically, energy may end up being the cause of the downfall of the society it helped us to build. What is most intriguing is the long history of development of renewable energy and alternative energy technologies and the concurrent growth of the fossil fuel industry. Over 100 years ago, engineers and scientists of the day questioned the viability of an economy based on nonrenewable energy and worried that the fossil fuel supply would be exhausted in the near future, causing untold hardships on nations that relied on them as a primary energy source. Historically, fossil fuel companies have used their political power to subvert the development of renewable and alternative energy sources.

5–8 million years ago   The evolution of modern humans from ancient human species like the now extinct *Homo erectus* begins in Africa. The only energy used at this time comes from the sun. Energy from the sun is used for light, heat, and, through the consumption of vegetation where

125

5–8 million years ago *(cont.)*	solar energy is converted and stored as potential energy, sustenance. The consumption of plants by humans allows this potential energy to be converted. As humans develop the skills to make tools and use them, their diet expands to include animals (Cook 1971, 135).
100,000– 400,000 years ago	*Homo erectus* discovers fire and begins using it for cooking, heating, and lighting. The main fuel for fires during this time is wood. This discovery enables humans to survive the Ice Age and continue evolving into *Homo sapiens,* the same species as modern-day humans. Fire becomes the first source of energy to be used for residential purposes (Cook 1971, 135).
40,000 years ago	*Homo sapiens* (modern humans) emerge and begin migrating around the globe.
10,000 years ago	Humans shift from hunter-gatherer to primitive agrarian or agriculturally based society. Energy usage increases at least sixfold as animals are domesticated and used for agricultural purposes. Power is derived from wind and water for a variety of purposes, although official documentations of this did not begin until thousands of years later.
5,000 years ago	Some experts believe that the use of wind and water for grinding grain begins in Egypt or Greece, though there are no official records to prove this (Iowa Energy Center 2006).
1200 BCE	Polynesians use sails to power ships with wind.
1000 BCE	China is the first country to use coal instead of wood for cooking and heating purposes (World Coal Institute 2005, 19).
Ca. 500 BCE	Thales, a Greek philosopher, discovers the existence of electricity by creating and observing static (Anzovin and Podell 2000, 91).
85 BCE	Antipater, a Greek poet, makes what is thought to be the first reference to the use of water mills in the

Western world, though it is widely accepted that they were used in Asia (Reynolds 1970, 11).

7 BCE According to the U.S. Department of Energy, "a magnifying glass is used to concentrate the sun's rays on a fuel and light a fire for light, warmth, and cooking" ("Solar History Timeline: 1900s" 2005).

3 BCE According to the U.S. Department of Energy, "Greeks and Romans use mirrors to light torches for religious purposes" ("Solar History Timeline: 1900s" 2005).

1 CE Hero of Alexandria is the first to document the use of a windmill, which powers an organ (Drachmann 1961, 141).

According to the U.S. Department of Energy, "passive solar design is incorporated into the design of Roman bath houses where large, south-facing windows were built to let in the sun's warmth" ("Solar History Timeline: 1900s" 2005).

6 According to the U.S. Department of Energy, passive solar design becomes so common that "the Justinian Code establishes 'sun rights' to ensure that a building has access to the sun" ("Solar History Timeline: 1900s" 2005).

9 An Islamic geographer records the use of vertical windmills in Persia (Hassan and Hill 1986, 54). Some experts believe that horizontal windmills are invented concurrently in Europe, perhaps in Germany (Preston 1923, 14).

1003 The Chinese drill natural gas wells and use bamboo pipes to vent the gas and fuel kilns and forge fires (Anzovin and Podell 2000, 92).

1100 A tidal-powered mill is built in England (Anzovin and Podell 2000, 89).

1180 The first horizontal axis windmills in Europe are documented (Preston 1923, 14).

1275   Marco Polo returns to Italy after exploring China to introduce coal as an energy source to the entire Western world.

1599   The first greenhouses to resemble those of modern times are used in Europe to protect the exotic plants that explorers are bringing back from foreign lands. This marks a major progression in the use of passive solar energy.

1663   Otto Von Guericke, a German physicist and engineer, invents the electric generator. Guericke becomes the first person to create and observe artificial lighting, albeit on a small, impractical scale (Anzovin and Podell 2000, 91). This development is the first of many that would have major consequences for the future of energy consumption.

1673   Father Louis Hennepin discovers coal in what would later become the state of Illinois (Kane, Anzovin, and Podell 1997, 222).

1698   Captain Thomas Savery invents the first steam engine for the express purpose of drawing water out of coal mines so that mining operations could proceed. Ironically, the most energy-intensive invention of the century is created to aid in the extraction of energy sources.

1700s   Coal becomes the dominant source of energy in Europe after forests are overharvested for their wood and the steam engine is put to use.

1712   Thomas Newcomen invents a more efficient steam engine than the Savery engine to draw water out of coal mines faster.

1729   Stephen Grey, an English inventor, achieves the conduction of electricity along a wire by rubbing his hands on a corked glass tube. The charge was transmitted almost 500 feet (Anzovin and Podell 2000, 91).

1750   Ben Franklin determines the nature of electricity while observing lightning. This development is critical for future developments in renewable energy technology.

1767   According to the U.S. Department of Energy, the Swiss scientist Horace de Saussure builds the world's first solar collector, which is later used by Sir John Herschel to cook food during his South African expedition in the 1830s ("Solar History Timeline: 1900s" 2005).

1769   James Watt patents his steam engine that is several times more powerful than the Savery-Newcomen engines and that can be used for applications other than in coal mining. This single invention is believed by many to be one of the most important turning points in the course of modern history. The Watt steam engine is widely responsible for the Industrial Revolution, which causes human consumption of energy to skyrocket. In a sense, Watt's steam engine is responsible for the energy situation we now find ourselves in (Routledge 1989, 4).

1774   Antoine Lavosier, a French chemist, develops the first solar furnace, considered by many to be the first modern form of solar energy technology.

1775   The U.S. Army Corps of Engineers is founded. The Corps becomes responsible for the design and construction of hundreds of hydroelectric projects in the United States.

1800   William Nicholson and Sir Anthony Carlisle, English scientists, are credited for being the first to use electrolysis to break water down into hydrogen and oxygen.

1802   Humphry Davy, an English chemist, describes the first fuel cell, which becomes the basis for generating electricity through electrolysis in future hydrogen energy applications (Anzovin and Podell 2000, 89).

1803   James Watt's foundry in Birmingham, England, is the first large building to be illuminated using coal gas as the fuel (Anzovin and Podell 2000, 93).

1804   Watt's steam engine is used in the invention of the locomotive. The use of engines instead of manpower or horsepower for transportation creates even more demand for energy from coal and wood, further increasing our consumption of these finite resources (Routledge 1989, 14).

1807   Watt's steam engine is incorporated into nautical designs, resulting in the invention of the first steamboat (Routledge 1989, 92).

1812   The London Gas Light and Coke Company becomes the first public power utility.

1816   According to the U.S. Department of Energy, "On September 27, 1816, Robert Stirling applies for a patent for his *economiser* at the Chancery in Edinburgh, Scotland. This engine is later used in the dish/Stirling system, a solar thermal electric technology that concentrates the sun's thermal energy to produce electric power" ("Solar History Timeline: 1900s" 2005).

1820   Reverend W. Cecil reads his paper, "On the Application of Hydrogen Gas to Produce Moving Power in Machinery," at Cambridge University. It is the first documented proposal for the use of hydrogen as a fuel (Hoffman 2001, 27).

1827   Benoit Fourneyron, a French engineer, invents the first water turbine not only that is more efficient than traditional waterwheels but that can endure more pressure like that produced by a hydroelectric dam (Anzovin and Podell 2000, 89).

1839   William R. Grove, an English chemist, invents the first fuel cell that is actually used to convert hydrogen into electricity (Anzovin and Podell 2000, 90).

According to the U.S. Department of Energy, "French scientist Edmond Becquerel discovers the photovoltaic effect while experimenting with an electrolytic cell made up of two metal electrodes placed in an electricity-conducting solution; the electricity generation increases when exposed to light" ("Solar History Timeline: 1900s" 2005).

1841   Frederick de Moleyns patents the first incandescent lamp in Great Britain. This bulb does not have a filament, like the later model of Thomas Edison and Joseph Swan, but relies on carbon powder heated between two platinum wires to produce light (Anzovin and Podell 2000, 92).

1850   The most popular lamp fuel is made from turpentine (distilled from wood), alcohol (distilled from grain), and camphor oil. Basically, biofuels are used widely before oil is discovered (Kovarik 2007).

1855   American Dr. Abraham Gesner patents a process for creating kerosene. Kerosene becomes an important fuel for lighting in the 19th century (Anzovin and Podell 2000, 93).

1857   The first biogas plant is built in Bombay, India, to provide methane fuel for a nearby leper colony (Anzovin and Podell 2000, 90).

1859   Edwin Drake strikes oil in Titusville, Pennsylvania, creating the first commercially productive oil well and setting off the oil drilling craze. Oil quickly becomes the preferred alternative to coal for lighting. Oil is used mainly in the form of kerosene (Melosi 1985).

Gaston Planté invents the first practical dry cell battery. This battery is the precursor to the modern automobile battery (Anzovin and Podell 2000, 91).

1860   Auguste Mouchout of France develops the first solar engine, which uses concentrated solar radiation to

1860 *(cont.)*	boil water and run an engine with the steam it produces (Smith 1995).
1869	Aristide Bergés of France invents the hydroelectric generator to supply electricity to the machines in his workshop (Anzovin and Podell 2000, 92).
1870	The oil industry is emerging, and the transition from coal to oil is underway (Melosi 1985, 39).
1872	John Ericsson invents the first solar parabolic trough (Smith 1995).
1873	According to the U.S. Department of Energy, Willoughby Smith discovers the photoconductivity of selenium ("Solar History Timeline: 1900s" 2005).
1876	Pavel Nikolayevich Yablochkov, a Russian electrical engineer, develops the first electric rail and street lamps, which are used in Russia and Europe for street lighting until improved technology emerges at the turn of the century (Anzovin and Podell 2000, 92).
	According to the U.S. Department of Energy, "William Grylls Adams and Richard Evans Day discover that selenium produces electricity when exposed to light. Although selenium solar cells fail to convert enough sunlight to power electrical equipment, they prove that a solid material can change light into electricity without heat or moving parts" ("Solar History Timeline: 1900s" 2005).
1878	William Adams begins construction on the first solar tower. His design is the basis for modern-day solar towers (Smith 1995).
	Edison Electric Light Company of New York becomes the first electric company.
1879	American inventor Thomas Edison and British inventor Joseph Swan merge patents for the first incandescent light bulb to use filaments to produce light in a

vacuum inside a glass bulb. This invention becomes the basis for modern incandescent light bulbs.

Rookes Evelyn Bell Crompton, an English electrical pioneer, becomes the first person to have a private home fully illuminated with electric lighting (Anzovin and Podell 2000, 92).

The first electricity to be generated by the water power of Niagara Falls lights up 16 lamps in Prospect Park, Niagara Falls, New York. Large-scale electricity production does not occur until 1890.

1880   Michigan's Grand Rapids Electric Light and Power Company builds the first hydroelectric power plant in the United States to provide lighting services. The plant uses a waterwheel to supply lighting power to the Wolverine Chair Company.

1882   The first hydroelectric power plant to provide incandescent lighting in the United States is built in Appleton, Wisconsin, to provide 12.5 kilowatts of power for two paper mills and a home.

1883   Charles Fritts, an American inventor, develops the first solar cell. It uses selenium, which is very inefficient, converting less than 1 percent of the energy it collected from the sun into electricity (Anzovin and Podell 2000, 90).

Marcel Deprez, a French inventor, conducts the first transmission of electricity through a high-voltage cable in Paris, France (Anzovin and Podell 2000, 92).

1885   Charles Tellier invents the first solar-powered water pump. It is the first solar-powered technology developed that does not use reflection as a technique for concentrating solar power (Smith 1995).

1886   Forty-five water-powered electric plants are now in operation in the United States and Canada.

1887   The first fully operational hydroelectric plant in the West comes online in San Bernardino, California. The plant is able to provide for more than just lighting, unlike the Grand Rapids Electric Light and Power Company plant in Michigan.

1888   According to the Energy Information Administration (EIA), "Charles F. Brush used the first large windmill to generate electricity in Cleveland, Ohio. The windmill starts to be called 'wind turbine.' In later years, General Electric acquired Brush's company, Brush Electric Co." (EIA 2005).

        The development of steel blades makes windmills more efficient. Six million windmills spring up across the United States as settlers moved west.

1889   The first hydroelectric power plant to generate electricity for use over a long distance is built by the Willamette Falls Electric Company at Willamette Falls, Oregon City, Oregon. At this time, 200 electric plants in the United States rely on water power for some or all power generation.

1890   The mass production of automobiles begins driving up the demand for oil.

1891   Indiana enacts a natural gas conservation law to prevent the use of the resource in wasteful applications (Kane, Anzovin, and Podell 1997, 224).

        The Baltimore inventor Clarence Kemp patents the first commercial solar water heater, which is still used extensively throughout the world ("Solar History Timeline: 1900s" 2005).

1892   According to the U.S. Department of Energy, "the first district heating system to be heated by geothermal energy is developed in Boise, Idaho" ("Solar History Timeline: 1900s" 2005).

1896   The first hydroelectric plant to use a storage battery is built by the Hartford Electric Light Company in Hartford, Connecticut. This makes it possible to store energy from water power that would have gone to waste during periods of low demand.

1898   The first underground hydroelectric plant is built behind the Snoqualmie Falls near Seattle, Washington (Anzovin and Podell 2000, 92).

1901   The Federal Water Power Act is established to govern the development of water power resources in the United States.

1902   The U.S. Bureau of Reclamation is established. The Bureau of Reclamation is responsible for commissioning hundreds of federal dams in the United States.

1904   The first geothermal plant goes into operation in Larderello, Italy.

Willhelm Halwachs discovers the photosensitive nature of copper and coprous oxide when combined.

Swedish scientist Svante Arrhenius conducts the first studies on the influence of increased concentrations of carbon dioxide in the atmosphere. Although his research recognizes a link between anthropogenic carbon emissions and warming, his conclusions are that the effects of this warming are beneficial for plant growth and food production. At this time, there are no suspicions in the scientific community that humans could actually cause dramatic changes in the climate.

1905   Albert Einstein publishes a paper on the photoelectric effect, also called the photovoltaic effect. His description of how light can be transformed into electricity earns him the Nobel Prize in physics in 1921.

1906   Taxes are repealed on industrial and fuel alcohol to encourage the use of ethanol. At this time, Europe is using ethanol to power cars and household appliances (Kovarik 2007).

1907   Hydropower provides 15 percent of the electricity-generating capacity in the United States.

1908   According to the U.S. Department of Energy, "William J. Bailey of the Carnegie Steel Company invented a solar collector using copper wires and an insulated box. The design is still used today for solar collectors" ("Solar History Timeline: 1900s" 2005). Bailey's company, Day and Night, is put out of business by 1920 due to discoveries of cheap natural gas (Kovarik 2007).

1909   Minidoka Dam on the Snake River in Idaho is the first dam to be built by the U.S. Bureau of Reclamation, Department of the Interior.

1910   Georges Claude, a French engineer and chemist, invents the neon light. It is used to illuminate the Grand Palais in Paris (Anzovin and Podell 2000, 92).

1914   Thomas Alva Edison invents the first alkaline dry cell battery. He does not commercialize it, however, and it does not become widely used until the 1970s (Anzovin and Podell 2000, 92).

The controversial O'Shaughnessy Dam is built in Hetch Hetchy Valley, the sister valley to Yosemite, flooding what environmentalists like John Muir deem a national treasure (Righter 2005).

1917   Alexander Graham Bell promotes the use of alcohol as a fuel in an issue of *National Geographic* (Kovarik 2007).

1920   Hydropower generates 25 percent of U.S. electricity.

An article in *Scientific American* declares that the majority of people are of the opinion that alcohol (ethanol) will be the fuel of the future (Kovarik 2007).

1921   According to the U.S. Department of Energy, "John D. Grant builds the United States' first geothermal power plant. The operation produced 250 kilowatts, enough electricity to light the buildings and streets at Grant's resort. The plant, however, is not competitive with other sources of power, and it soon falls into disuse" ("Solar History Timeline: 1900s" 2005).

1923   Leaded gasoline goes on sale for the first time. The lead replaces ethanol as a gasoline additive even though flex-fuel engines were created to run on the latter (Kovarik 2007).

1925   Henry Ford promotes the use of ethanol, which he deems the "fuel of the future," for fueling automobiles (Kovarik 2007).

1927   The first hydroelectric plant to use water from a reservoir is constructed by the Connecticut Light and Power Company at Rocky River, Connecticut. Water stored in a reservoir is used to generate electricity in a 33,000-horsepower turbine.

1928   The St. Francis Dam in California bursts, killing 511 people on March 12, showing that hydroelectric dams are not guaranteed to be safe (Kovarik 2007).

1930   The first commercial greenhouse use of geothermal energy is undertaken in Boise, Idaho.

        According to the U.S. Department of Energy, "In Klamath Falls, Charlie Lieb develops the first downhole heat exchanger (DHE) to heat his house. Today, more than 500 DHEs are in use around the country" ("A History of Geothermal Energy in the United States" 2006).

1931   The U.S. Bureau of Reclamation of the Department of the Interior awards a contract to build the first concrete arch-gravity dam in Boulder City, Nevada. The dam's turbines begin producing power in 1936. Originally named the Boulder Dam, it is renamed Hoover Dam in 1947. It is one of the biggest dams in the world.

1933   The Tennessee Valley Authority (TVA) is established by the U.S. federal government. The TVA is created as a government-owned electricity provider to help the Tennessee Valley, which was severely affected by the Great Depression. The TVA is responsible for the construction of several dams in the area.

1934   Arthur Compton, an American electrical engineer and inventor at the General Electric Company, invents the first fluorescent light bulb.

        The first international oil pipeline is completed, running from the oil fields of northern Iraq to Syria (Anzovin and Podell 2000, 93).

1936   Leo Szilard, a Hungarian American physicist, discovers the nuclear chain reaction necessary for nuclear fission.

1937   The Hindenburg disaster results in a debilitating stigma for hydrogen despite the fact that not one of the 36 deaths is caused by the burning hydrogen. Thirty-four people die from sustained injuries after jumping from the blimp, and two die of burns caused by flaming on-board diesel fuel.

1938   The first government-financed ethanol plant is opened in Atchison, Kansas, but the plant is not a commercial success (Kovarik 2007).

1939   Electric power generated by cosmic rays is carried by wire to the New York World's Fair to power the lights at the Lagoon of the Nations.

1940   Hydropower generates 40 percent of U.S. electricity.

1941   Russell Ohl, an American, is the first to replace selenium in solar cells with silicon, which proves to be slightly more efficient than Fritts's original design (Anzovin and Podell 2000, 90).

        A wind turbine invented by Palmer Cosslet Putnam generates electricity in Vermont, becoming the first turbine in the United States to be connected to a central power system for the purpose of energy production.

1942   Enrico Fermi builds the first nuclear reactor while working on the Manhattan Project at the University of Chicago.

1943   Boulder Dam, later renamed Hoover Dam, becomes the first hydroelectric power plant to produce a million kilowatts of electricity.

1946   The first commercially sized solar furnace is built in Meudon, France (Anzovin and Podell 2000, 90).

         The first Soviet nuclear reactor goes online and begins producing power. The F-1, as the plant is named, is designed by a team of Soviet physicists under Igor Vasilyevich Kurchatov. Initially, it produces 10 watts of power (Anzovin and Podell 2000, 94).

1947   According to the U.S. Department of Energy, "demand for passive solar building designs increases in the U.S. in response to the energy shortages caused by the war" ("Solar History Timeline: 1900s" 2005).

1948   Geothermal technology moves east when Professor Carl Nielsen of Ohio State University develops the first ground-source heat pump, for use at his residence ("A History of Geothermal Energy in the United States" 2006).

         According to the U.S. Department of Energy, "J. D. Krocker, an engineer in Portland, Oregon, pioneers the first commercial building use of a geothermal groundwater heat pump" ("A History of Geothermal Energy in the United States" 2006).

1951   Electric power from a fast breeder nuclear reactor is obtained in the United States for the first time in Idaho Falls, Idaho.

1954   Soviet engineers build the first civilian power plant outside Moscow. It produces 5 megawatts of power (Anzovin and Podell 2000, 94).

Bell Labs creates the first solar photovoltaic cell to produce "useful" power, converting 6 percent of the sunlight hitting the cell into electricity. Three American researchers, Gerald L. Pearson, Daryl Chapin, and Calvin Fuller, are responsible for its development (Anzovin and Podell 2000, 90).

1955   Electric power from nuclear energy is sold commercially in the United States for the first time by the Atomic Energy Commission to the Niagara Mohawk Power Corporation in Milton, New York.

The first solar cell to be produced commercially in the United States is made by National Fabricated Products, a Chicago-based company.

According to the U.S. Department of Energy, "U.S. architect Frank Bridgers designs the world's first commercial office building to use solar water heating and passive solar design" ("Solar History Timeline: 1900s" 2005).

1956   The first nuclear power plant to produce a substantial amount of electricity is Calder Hill in England.

The first Asian nuclear power plant, Apsara, goes online in Trombay, India. It is designed and built by the Atomic Energy Establishment, Trombay, founded by Indian physicist and mathematician Homi Jehangir Bhabha.

1957   According to the U.S. Department of Energy, "Hoffman Electronics develops a solar cell with 8 percent efficiency. Just one year later they increase efficiency to 9 percent" ("Solar History Timeline: 1900s" 2005).

Roger Revelle coauthors a paper with Hans Suess suggesting that human-induced gas emissions might create a "greenhouse effect" and cause global warming over time. This is the first paper to seriously consider the effects of fossil fuel use on the climate.

1959   The first experimental fast breeder reactor, the United Kingdom Atomic Energy Authority Technology Prototype Fast Reactor, is built in Dounreay, Scotland. It is closed in March 1994 (Anzovin and Podell 2000, 94).

Harry Karl Ihrig demonstrates the first hydrogen fuel cell to power a vehicle—a 20-horsepower tractor (Hydrogen Now 2001).

The Malpasset Dam in France bursts, killing 420 people (Kovarik 2007).

1960   According to the U.S. Department of Energy, Hoffman Electronics develops a solar cell that reaches 14 percent energy conversion efficiency, a 5 percent increase from 1958 ("Solar History Timeline: 1900s" 2005).

The first experimental solar thermal power plant is finished in what was then the capital of Ashkhabad (now Ashgabat, Turkmenistan). It has more than 1,200 mirrors that concentrate sunlight on a central steam boiler, which is connected to an electric generator (Anzovin and Podell 2000, 90).

The first nuclear power plant to operate commercially in the United States is built by the Yankee Atomic Electric Company in Rowe, Massachusetts. It is permanently shut down for safety reasons in 1995.

According to the U.S. Department of Energy, "the country's first large-scale geothermal electricity-generating plant begins operation. Pacific Gas and Electric operates the plant, located at The Geysers in California. The first turbine produces 11 megawatts (MW) of net power and operates successfully for more than 30 years. Today, 69 generating facilities are

1960 *(cont.)*	in operation at 18 resource sites around the country" ("A History of Geothermal Energy in the United States" 2006).

1961   Construction of the Rance River Power station, the first commercial tidal electric generator plant, begins in Brittany, France, in the estuary of the Rance River. The plant is completed in 1967. The design includes a dam with turbines that are reversible so that energy can be produced whether the tide is coming in or going out. The plant is capable of producing 240,000 watts of electricity (Anzovin and Podell 2000, 90).

Three workers are killed by high levels of radiation from a federally supported experimental nuclear reactor in Idaho Falls, Idaho. These are the first fatalities to be attributed to nuclear energy in the United States.

1969   The Tarapur Atomic Power Station in Maharashtra becomes India's first nuclear plant for commercial power production. It produces 420 megawatts of electricity.

On January 28, an offshore oil rig begins leaking crude oil off the coast of Santa Barbara. The spill has disastrous effects on wildlife and on more than 200 miles of beaches, causing a stir among the American public about the environmental risks of oil exploration.

According to the U.S. Department of Energy, "a solar furnace is constructed in Odeillo, France featuring an eight story parabolic mirror" ("Solar History Timeline: 1900s" 2005).

1970   According to the U.S. Department of Energy, "Dr. Elliot Berman designs a solar cell that brings the cost down from $100 per watt to $20 per watt, prompting a growth in domestic use of solar powered applications" ("Solar History Timeline: 1900s" 2005).

Construction of the Itaipu Dam on the border of Brazil and Paraguay begins. Twenty-five thousand

Guarani Indians are displaced, and people around the world begin questioning the value of massive hydro-electric dam projects (Kovarik 2007).

According to the U.S. Department of Energy, "the Geothermal Resources Council is founded to encourage development of geothermal resources worldwide" ("A History of Geothermal Energy in the United States" 2006).

According to the U.S. Department of Energy, "the Geothermal Steam Act is enacted, which provides the Secretary of the Interior with the authority to lease public lands and other federal lands for geothermal exploration and development in an environmentally sound manner" ("A History of Geothermal Energy in the United States" 2006).

1972   The University of Delaware creates the Institute of Energy Conversion, the world's first laboratory dedicated to research and development in the area of thin-film photovoltaic and solar thermal systems ("Solar History Timeline: 1900s" 2005).

The Buffalo Creek Dam in West Virginia bursts, killing 125 people (Kovarik 2007).

According to the U.S. Department of Energy, the Geothermal Energy Association is formed. The association includes U.S. companies that develop geothermal resources worldwide for electrical power generation and direct heat uses ("A History of Geothermal Energy in the United States" 2006).

1973   The oil crisis of 1973 begins on October 16 when an Arab oil embargo is imposed on countries that support Israel in the Yom Kippur War. In addition, the Organization of the Petroleum Exporting Countries (OPEC) agrees to raise the price of oil, taking advantage of the dependency of industrialized nations on their oil supplies. This results in shortages and inflation in countries dependent on the importation of oil

1973 *(cont.)*	from these countries. The United States is especially hard hit.
1974	Government-sponsored research into wave power begins in earnest in response to the oil crisis caused by the Arab embargo of 1973 and the excessive dependency of industrialized countries on imported oil. Japan, Norway, and the United Kingdom are instrumental in advancing this technology. All these programs fail, however, to achieve market penetration of wave technologies, damaging the credibility of this energy source.

According to the U.S. Department of Energy, the National Aeronautics and Space Administration (NASA) develops a two-bladed wind turbine at the Lewis Research Center in Cleveland, Ohio, in response to the oil crisis ("Solar History Timeline: 1900s" 2005). |
| 1975 | The Shimantan Dam in China bursts, killing no fewer than 250,000 people (Kovarik 2007). |
| 1976 | The Grand Teton Dam in Idaho bursts, killing 14 people (Kovarik 2007). |
| 1977 | A 2-megawatt windmill is installed at Howard's Knob in Boone, North Carolina, by General Electric. It is dismantled in 1983 following local residents' complaints of noise (Kovarik 2007).

The U.S. Department of Energy (DOE) is formed. |
| 1978 | The Carter administration enacts the first electricity feed-in law in the world under the Public Utility Regulatory Policies Act (PURPA) to promote the use of renewable energy in the United States. The electricity feed in-law, known in some countries as the advanced renewable energy tariff, requires utility companies to buy a percentage of their electricity from nonutility power producers. Electricity feed-in laws are deemed by many to be the most successful policy mechanism for increasing the share of renewable energy in the energy economy (Chapters 2 and 3). |

1979   A partial core meltdown caused by equipment failures and human error at the Three Mile Island nuclear reactor in Middleton, Pennsylvania, on March 28 results in the evacuation of nearby neighborhoods. This accident prompts a speedy and forceful rejection of all new nuclear projects by the American public, a sentiment that continues into the 21st century.

After the energy crisis, proposals for ethanol-blended gasoline reemerge. (At this time, it seems like a new idea, as it has not been explored since the 1930s and 1940s.) Oil industries lobby against such proposals (Kovarik 2007).

In response to the energy crisis, Brazil begins investing heavily in ethanol production from sugarcane. By 2007, over 40 percent of Brazil's automobiles are powered by ethanol, making the country practically independent of imported oil (Kovarik 2007).

According to the U.S. Department of Energy, "the first wind turbine rated over 1 megawatt (MOD-1), begins operating. It had a 2-megawatt capacity rating" ("Solar History Timeline: 1900s" 2005).

1980   A photovoltaic power plant, made up of 266,029 solar cells, is dedicated on June 7 to Governor Scott Matheson at the Natural Bridges National Monument, Utah. The plant provides enough power for six staff residences, maintenance facilities, a water sanitation system, and a visitors' center. The U.S. National Park System, the U.S. Department of Energy, and the Massachusetts Institute of Technology's Lincoln Laboratory collaborate on the project (Anzovin and Podell 2000, 90).

According to the U.S. Department of Energy, "the Crude Oil Windfall Profits Tax Act further increased tax credits for businesses using renewable energy. The Federal tax credit for wind energy reached 25 percent and rewarded businesses choosing to use renewable energy" ("Solar History Timeline: 1900s" 2005).

1981   According to the U.S. Department of Energy, "Paul MacCready builds and flies the first solar powered aircraft from France to England across the English Channel. The aircraft had more than 16,000 solar cells mounted to the wings to produce over 3000 watts of power" ("Solar History Timeline: 1900s" 2005).

1982   Solar One, a 10-megawatt solar power tower plant, begins operating in California. The plant stops operating in 1988 (Chapter 1).

According to the U.S. Department of Energy, "worldwide, photovoltaic production exceeds 9.3 megawatts" ("Solar History Timeline: 1900s" 2005).

1983   According to the U.S. Department of Energy, "Solar Design Associates completes a home powered by an integrated, stand-alone, 4-kilowatt photovoltaic system in the Hudson River Valley" ("Solar History Timeline: 1900s" 2005).

According to the U.S. Department of Energy, "worldwide, photovoltaic production exceeds 21.3 megawatts" ("Solar History Timeline: 1900s" 2005).

According to the U.S. Department of Energy, "because of a need for more electricity, California utilities [contract] with facilities that qualified under PURPA to generate electricity independently. The price set in these contracts [is] based on the costs saved by not building the planned coal plants" ("Solar History Timeline: 1900s" 2005).

1984   Luz International Ltd. begins construction on the first of nine commercial-scale thermal solar energy installations in the United States. Each unit produces 30,000 kilowatts, and, by 1990 when the final plant is built, the cost per kilowatt is reduced by 50 percent. The company goes bankrupt while building the tenth plant due to the falling price of natural gas and oil (Anzovin and Podell 2000, 90).

According to the U.S. Department of Energy, "the Sacramento Municipal Utility District commissions its first 1-megawatt photovoltaic electricity generating facility" ("Solar History Timeline: 1900s" 2005).

Superphénix, the first commercially sized fast breeder reactor, is built in France. Technological problems prevent the plant from producing electricity for more than nine months over a 10-year period. French prime minister Lionel Jospin orders the Superphénix closed in 1997 after it experiences a brief period of productivity from late 1995 to late 1996 (Anzovin and Podell 2000, 95).

1985   According to the U.S. Department of Energy, "researchers at the University of South Wales break the 20% efficiency barrier for silicon solar cells" ("Solar History Timeline: 1900s" 2005).

Fifteen alcohol fuel plants are producing 50 million gallons of ethanol per year in Virginia. By 1992, every one of these plants is bankrupt (Kovarik 2007).

According to the U.S. Department of Energy, "California wind capacity exceeds 1,000 megawatt, enough power to supply 250,000 homes. But the wind turbines used [are] very inefficient resulting in their dismantling and replacement in the late 1980s" ("Solar History Timeline: 1900s" 2005).

1987   The Sanyo Corporation of Japan develops the first thin-film solar cells. The company replaces the expensive and difficult-to-manufacture crystalline silicon material with a less expensive amorphous (noncrystalline) silicon (Anzovin and Podell 2000, 90).

1988   A temperature of 100 million Kelvin is achieved. This is considered the minimum temperature necessary to sustain a controlled fusion reaction.

1990   According to the U.S. Department of Energy, "more than 2,200 megawatts of wind energy capacity [is]

1990 *(cont.)*	installed in California—more than half of the world's capacity at the time" ("Solar History Timeline: 1900s" 2005).
1991	President George H. W. Bush announces that the U.S. Department of Energy's Solar Energy Research Institute has been designated the National Renewable Energy Laboratory ("Solar History Timeline: 1900s" 2005).
1992	Sanyo Electric and Hyogo Prefecture of Japan develop the first practical, portable fuel cell. The cell is capable of producing 250 watts of electricity (Anzovin and Podell 2000, 91).
	According to the U.S. Department of Energy, "researchers at the University of South Florida develop a 15.9% efficient thin-film photovoltaic cell made of cadmium telluride, breaking the 15% barrier for this technology" ("Solar History Timeline: 1900s" 2005).
	According to the U.S. Department of Energy, the Energy Policy Act is passed. The act reforms the Public Utility Holding Company Act and many other laws dealing with the electric utility industry. It also authorizes a production tax credit of 1.5 cents per kilowatt hour for wind-generated electricity ("Solar History Timeline: 1900s" 2005).
1993	The first Green Power Market Program is started in the United States.
	A commercial electric plant powered by poultry litter begins operating in Suffolk, England. It is designed by Lifschutz Davidson (Anzovin and Podell 2000, 90).
	Intersource Technologies, a California-based company, develops the E-lamp. The E-lamp is an electronic light bulb that uses 75 percent less electricity than an incandescent light bulb but with the same brightness. The E-lamp is a breakthrough in energy conservation technology (Anzovin and Podell 2000, 93).

According to the U.S. Department of Energy, "Pacific Gas & Electric installs the first grid-supported photovoltaic system in Kerman, California. The 500-kilowatt system is the first 'distributed power' PV installation" ("Solar History Timeline: 1900s" 2005).

According to the U.S. Department of Energy, "the National Renewable Energy Laboratory . . . completes construction of its Solar Energy Research Facility; it will be recognized as the most energy-efficient of all U.S. government buildings in the world" ("Solar History Timeline: 1900s" 2005).

1994   The Tokamak Fusion Test Reactor at the Plasma Physics Laboratory of Princeton University becomes the first fusion reactor to generate 10 megawatts of power. This is considered to be an important milestone by fusion power researchers (Anzovin and Podell 2000, 95).

According to the U.S. Department of Energy, "the National Renewable Energy Laboratory develops a solar cell made of gallium indium phosphide and gallium arsenide; it's the first one of its kind to exceed 30% conversion efficiency" ("Solar History Timeline: 1900s" 2005).

Construction of the Three Gorges Dam in China begins. The dam becomes the largest hydroelectric dam on the planet and the most controversial, with over 1.13 million people displaced (Kovarik 2007).

1995   The first commercial wave-powered electricity generator, the OSPREY (Ocean Swell Powered Renewable Energy), is anchored to the seabed and begins operating near the mouth of the River Clyde, Scotland. It is designed to generate 2 megawatts of electricity from wave power and another 1.5 megawatts from wind turbines attached at the surface. It is also the first plant to commercially power water desalinization plants (Anzovin and Podell 2000, 91).

1995
(cont.)

According to the U.S. Department of Energy, Kenetech, the producer of most of the U.S.-made wind generators, faces financial difficulties, sells off most of its assets, and stops making wind generators after the Standard Offer Number 4 contract rollovers in California lead to lower rates being paid to the Qualifying Facilities (QF). "This '11th-year cliff' create[s] financial hardships for most QFs on these contracts" ("Solar History Timeline: 1900s" 2005).

1996

According to the U.S. Department of Energy, "the world's most advanced solar-powered airplane, the Icare, flies over Germany. Its wings and tail surfaces are covered by 3,000 super efficient solar cells, for a total area of 21 square meters" ("Solar History Timeline: 1900s" 2005).

According to the U.S. Department of Energy, the department joins with an industry consortium to "begin operating Solar Two—an upgrade of the Solar One concentrating solar power tower. Until the project's end in 1999, Solar Two demonstrates how solar energy can be stored efficiently and economically so power is produced even when the sun isn't shining; it also spurs commercial interest in power towers" ("Solar History Timeline: 1900s" 2005).

1998

The Stellarator, an improved version of the Tokamak fusion reactor, is built by Japan's National Institute of Fusion Science.

According to the U.S. Department of Energy, "Subhendu Guha, a scientist noted for pioneering work in amorphous silicon, leads the invention of flexible solar shingles, a roofing material and state-of-the-art technology for converting sunlight to electricity on buildings" ("Solar History Timeline: 1900s" 2005).

1999

Four solar service stations for electric vehicles are opened in Germany and the Netherlands by Royal Dutch Shell. The purpose of the stations is to recharge electric vehicles with power generated from solar energy (Anzovin and Podell 2000, 91).

According to the U.S. Department of Energy, "construction is completed on 4 Times Square in New York, the tallest skyscraper built in the city in the 1990s. It has more energy-efficient features than any other commercial skyscraper and includes building-integrated photovoltaic (BIPV) panels on the 37th through the 43rd floors on the south- and west-facing facades to produce part of the building's power" ("Solar History Timeline: 1900s" 2005).

According to the U.S. Department of Energy, "Spectrolab, Inc., and the National Renewable Energy Laboratory develop a 32.3 percent efficient solar cell. The high efficiency results from combining three layers of photovoltaic materials into a single cell, which is most efficient and practical in devices with lenses or mirrors to concentrate the sunlight. The concentrator systems are mounted on trackers to keep them pointed toward the sun" ("Solar History Timeline: 1900s" 2005).

According to the U.S. Department of Energy, "researchers at the National Renewable Energy Laboratory develop a record-breaking prototype solar cell that measures 18.8 percent efficient, topping the previous record for thin-film cells by more than 1 percent" ("Solar History Timeline: 1900s" 2005).

According to the U.S. Department of Energy, "cumulative installed photovoltaic capacity reaches 1000 megawatts, worldwide" ("Solar History Timeline: 1900s" 2005).

2000   According to the U.S. Department of Energy, "First Solar begins production at the Perrysburg, Ohio, photovoltaic manufacturing plant, the world's largest at the time; estimates indicate that it can produce enough solar panels each year to generate 100 megawatts of power" ("Solar History Timeline: 1900s" 2005).

2000 *(cont.)*	Industry Researchers develops a new inverter for solar electric systems that increases safety during power outages.
	According to the U.S. Department of Energy, "installed capacity of wind-powered electricity generating equipment exceeded 2,500 megawatts. Contracts for new wind farms continue to be signed" ("Solar History Timeline: 1900s" 2005).
2001	According to the U.S. Department of Energy, "the National Space Development Agency of Japan, NASDA, announces plans to develop a satellite-based solar power system that beams energy back to Earth. A satellite with large solar panels would use laser technology to transmit solar power to an airship at an altitude of about 12 miles; the airship would then transmit power to Earth" ("Solar History Timeline: 1900s" 2005).
2002	According to the U.S. Department of Energy, "the largest solar power facility in the Northwest—the 38.7-kilowatt White Bluffs Solar Station—goes online in Richland, Washington" ("Solar History Timeline: 1900s" 2005).
2003	Seventy-two plants produce over 3 billion gallons of ethanol—enough to raise octane in nearly one-quarter of the nation's gasoline supply. By 2005, the total is closer to 6 billion gallons (Kovarik 2007).

References

Anzovin, Steven, and Janet Podell. *Famous First Facts: International Edition*. New York: H. W. Wilson Company, 2000.

Cook, E. "The Flow of Energy in an Industrial Society." *Scientific American* 225, 3 (September 1971): 135–144.

Drachmann, A. G. "Heron's Windmill." *Centaurus* 1, 7 (1961): 145–151.

Energy Information Administration (EIA). "Wind Power Timeline." Energy Kid's Page. September 2005. www.eia.doe.gov/kids/history/timelines/wind.html

Foley, Gerald. *The Energy Question*. New York: Viking Penguin, 1989.

Hassan, Ahmad Y., and Donald Routledge Hill. *Islamic Technology: An illustrated history*. New York: Cambridge University Press, 1986.

"A History of Geothermal Energy in the United States." U.S. Department of Energy, Energy Efficiency and Renewable Energy. November 1, 2006. www1.eere.energy.gov/geothermal/printable_versions/history.html

Hoffman, P. *Tomorrow's Energy: Hydrogen, Fuel Cells, and the Prospects for a Cleaner Planet*. Cambridge, MA: MIT Press, 2001.

Hydrogen Now. "History." 2001. www.hydrogennow.org/Facts/History .htm

Iowa Energy Center. "A History of Wind Energy." *Wind Energy Manual*. 2006. www.energy.iastate.edu/renewable/wind/wem/wem–04_history .html

Kane, Joseph, Steven Anzovin, and Janet Podell. *Famous First Facts: A Record of Famous First Happenings, Discoveries and Inventions in American History*, 5th ed. New York: H. W. Wilson, 1997.

Kovarik, Bill. "Renewable Energy History." 2007. www.radford.edu /~wkovarik/envhist/RenHist/intro.html

Melosi, Martin V. *Coping with Abundance: Energy and Environment in Industrial America*. New York: Alfred A. Knopf, 1985.

Preston, Hayter. *Windmills*. New York: Dodd, Mead and Company, 1923.

Reynolds, John. *Windmills and Watermills*. Westport, CT: Praeger, 1970.

Righter, Robert W. *The Battle over Hetch Hetchy*. New York: Oxford University Press, 2005.

Routledge, Robert. *Discoveries and Inventions of the Nineteenth Century*. New York: Crescent Books, 1989.

Smith, Charles. "Revisiting Solar Power's Past." *Technology Review* 8, 5 (July 1995): 38-47.

"Solar History Timeline: 1900s." U.S. Department of Energy, Energy Efficiency and Renewable Energy. 2005. www1.eere.energy.gov/solar /solar_time_1900.html

World Coal Institute. "How Coal Is Used." *The Coal Resource*. 2005. www .worldcoal.org/pages/content/index.asp?PageID=37

5

Biographical Sketches

The people profiled in this chapter have made an impact in the field of alternative and renewable energy. They have been chosen for their contributions as scientists, inventors, scholars, advocates, and entrepreneurs in energy-related disciplines. Of this group, many of the scientists, inventors, and scholars laid the groundwork for later developments, and their work is extremely important for making alternative and renewable energy a viable source of power. Others made the discoveries that highlight the need for diversifying our energy economy, such as the occurrence of global warming and its causes. The advocates and entrepreneurs are responsible for raising awareness of the need for alternative and renewable energy and how it can meet our energy needs. Some disseminate information about the effect of energy use on our environment and health, and they encourage the adoption of responsible policies that take these facts into account. All of these people are important parts of the alternative and renewable energy equation, especially in their contributions to the debate over how to move beyond carbon-based energy resources.

William Grylls Adams (1836–1915)

William Grylls Adams was an English scientist who taught as a professor in the Department of Natural Philosophy at King's College. He is notable for his contribution to the discovery of the photoelectric effect, on which all solar energy applications are based. He was inspired by Auguste Mouchout's invention of the solar steam engine. With the intent of making improvements to

Mouchout's design, Adams began to experiment with different materials and designs. In 1876, working in conjunction with his student, Richard Day, he discovered that selenium produced electricity when exposed to sunlight. Using the selenium, he then added mirrors to the design to concentrate sunlight on the engine. This design came to be known as the power tower concept and is still in use today. In 1878, Adams wrote the first book on solar energy, *Solar Heat: A Substitute for Fuel in Tropical Countries*, which argued that solar power should be used as a fuel in tropical countries. Later that year, he displayed his power tower solar engine invention before many English notables. The resulting consensus was that, while solar energy might not be able to replace coal and wood, it could feasibly be used as a supplementary source of power. Unfortunately, he halted experimentation after he proved the feasibility of his invention and ceased contributing to the development of solar energy. Virtually all large-scale, centralized solar plants that use towers rely on his design of concentrating sunlight with the use of mirrors onto a central engine.

Svante Arrhenius (1859–1927)

Svante Arrhenius was a Swedish chemist considered to be one of the founders of the science of physical chemistry. He attended the University of Uppsala from 1874 to 1881 and received a master's degree in physics from the Academy of Sciences in Stockholm in 1884. In 1903, he was the first Swedish scientist to be awarded the Nobel Prize in chemistry. He was the first to discover that carbon dioxide in the atmosphere created a greenhouse effect, which years later became the basis for explaining the phenomenon of global warming. He developed a theory to explain the Ice Ages, in which he determined that changes in the levels of carbon dioxide in the atmosphere had the potential to significantly affect the temperature through the greenhouse effect. Arrhenius made predictions about the exact range in temperature change that would occur, and his predictions are incredibly close to the most recent estimates of the International Panel on Climate Change (IPCC). Arrhenius, however, had based his estimates on the rate of expected carbon dioxide level rise on the emissions of his time. He estimated that the expected doubling of carbon dioxide would take about 3,000 years. Since then, industrial carbon dioxide levels have risen at a much faster rate, and it is now expected that the doubling will take about century. Arrhenius laid the groundwork for subsequent work on climate change, which became a major driving force for the transition to renewable energy.

Jacques-Arsène d'Arsonval (1851–1940)

Jacques-Arsène d'Arsonval was a French biophysicist. He was famous for his work in the field of electrophysiology, the study of the effects of electricity on living organisms, where he contributed several important inventions. He was the director of the laboratory of biological physics at the College de France as well as a professor. In 1881, he suggested that we could tap the thermal energy of the oceans to create electricity. He became the first person to conceive of ocean thermal energy conversion (OTEC), making him an important contributor to renewable energy technology. His idea was for a closed-cycle system, as discussed in Chapter 1, which has been used in most modern-day OTEC pilot plants. His student, Georges Claude, was the first person to build a prototype OTEC plant off the coast of Cuba in 1930. D'Arsonval is an important figure in renewable energy for his conception of OTEC and his work as an educator, which ultimately resulted in the advancement of renewable energy technology by his students.

Alexandre-Edmond Becquerel (1820–1891)

Alexandre-Edmond Becquerel was a French physicist known for his work on electricity. Becquerel's father, Antoine César Becquerel, was a well-respected physicist and his son, Henri Becquerel, won the Nobel Prize in 1903 for his discovery of radioactivity. His father was an experimental professor of the Musee d'Histoire Naturelle, where Becquerel served as his pupil, assistant, and successor. He eventually received his doctorate from the University of Paris in 1840. In 1839, when he was just 19 years old, he discovered the photovoltaic effect while experimenting with an electrolytic cell made up of two metal electrodes placed in an electricity-conducting solution. He observed that the electricity generation increased when the electrodes were exposed to light. Solar cells are based on the physics of the photovoltaic effect. His work is foundational for today's work on photovoltaic cells that are used to generate electricity from solar power.

Elliott Berman (b. 1944)

Elliott Berman, an American chemist, graduated from Brown University and received a PhD in chemistry from Boston University. His biggest contribution to the field of renewable energy is his development of a solar cell in 1970. His solar cell was able to drive down the cost of solar power from $100 per watt to $20 per watt, making it economically feasible for solar to be used in a variety of domestic applications. He is also the founder of the Solar Power Corporation, an early maker of solar cells in Braintree, Massachusetts. He currently works in New York as the vice president of the Zentox Corporation, an environmental cleanup company.

Homi Jehangir Bhabha (1909–1966)

Homi Jehangir Bhabha was an Indian nuclear physicist. He played a major role in the development of the Indian atomic energy program and is considered the father of India's nuclear program. After receiving his doctorate from the University of Cambridge in 1934, he went on to establish the Atomic Energy Commission (AEC) in India in 1948. The Atomic Energy Commission worked toward the peaceful development of atomic energy. India's first atomic power plant, the Tarapur Maharashtra, was built in 1963 and went online in 1969. In 1974, eight years after Bhabha died in a plane crash, Indian scientists working for the AEC exploded a nuclear device, making India the sixth country in the world to acquire nuclear weapons.

Frank Bridgers (1932–2005)

Frank Bridgers was a U.S. mechanical engineer and architect. He established the Albuquerque-based mechanical engineering firm of Bridgers & Paxton Consulting Engineers in 1951. He is well-known for designing the first commercial office building to use solar water heating and passive solar design in 1955. Known as the Solar Building, it is located at 213 Truman St., N.E., in Albuquerque, and is now designated a historic landmark. The Solar Building, the world's first commercial structure to use the heat of the sun as its primary source for interior heating, was featured in *Life* magazine in 1956 and earned Bridgers a place in the Solar Energy Hall of Fame. In 1990, he was awarded the American Society of Heating and Ventilating Engineers' highest recognition, the F. Paul Anderson Medal.

Gro Harlem Brundtland (b. 1939)

Dr. Gro Harlem Brundtland began her career as a physician and scientist in the Norwegian public health system. She became involved in politics at the young age of seven as a member of the Norwegian Labour Movement. As an adult she served in public office for over 20 years, 10 of which she spent as the prime minister of Norway. In 1976, Brundtland was appointed to the position of minister of Environment, where she used her medical knowledge to find links between health and the environment. In the 1980s, she served as the chair of the World Commission of Environment and Development, which published its famous report *Our Common Future*, known informally as the Brundtland Report, in 1987. The report served as the catalyst for the 1992 United Nations Conference on Environment and Development (UNCED), also known as the Earth Summit, in Rio de Janeiro. The Brundtland Report is famous for developing the principle of sustainable development, on which nations have been trying to base their energy policies, albeit unsuccessfully in most cases. Brundtland now serves as a special envoy on climate change for the United Nations Secretary-General, Ban Ki-moon, and continues to explore the capacity of sustainable development in providing the world's citizens with a higher standard of living.

Jimmy Carter (b. 1924)

Although he is known throughout the world as a renewable energy advocate, Jimmy Carter is best-known for serving as the 39th president of the United States from 1977 to 1981. As president in the midst of an oil shortage during the OPEC embargo,

Carter made a significant attempt to move the United States toward energy independence. While in office, he created the Department of Energy and established the first comprehensive national energy plan. The plan included taxes on oil company profits, taxes on vehicles that did not achieve a certain level of gas mileage, requirements for builders to use more insulation, energy efficiency improvements for household appliances, tax credits for homeowners who switched to solar-powered hot water heaters, and support for mass transit projects. He also promoted energy conservation by example. Carter promoted wearing sweaters and turning the heat down, installed solar hot water panels on the White House, and put temperature controls on government buildings to prevent people from turning thermostats up too high or down too low. He encouraged the development of renewable energy and was the first U.S. president to encourage ethanol research and development. Funding for renewable energy projects soared under Carter, although most of his efforts were killed off during President Ronald Reagan's term after he successfully campaigned against Carter in 1980.

Georges Claude (1870–1960)

Georges Claude was a French engineer and chemist who is best-known for inventing the modern neon light in 1910, which he then introduced in the United States. His contribution to the renewable energy field is in the building of the first ocean thermal energy conversion plant. He was also a student of Jacques-Arsène d'Arsonval, the inventor of the OTEC concept. Claude built the first prototype OTEC plant off the coast of Cuba in 1930 and was able to produce 22 kilowatts of energy. He built a second plant off the coast of Brazil aboard a 10,000-ton cargo ship. Although both plants were destroyed by weather before they could become net energy producers, the designs provided the foundation for future OTEC energy developments.

Arthur Compton (1892–1962)

Arthur Compton was an American physicist and inventor at the General Electric Company. He received his PhD from the University of Wooster, where the subject of his dissertation was X-rays.

In 1927, he was awarded the Nobel Prize for his discovery of the Compton effect, which describes the behavior of X-rays when they interact with matter. His research findings had a powerful effect on quantum physics and ultimately contributed to the invention of the first nuclear reaction achieved by the Manhattan Project at the University of Chicago. The Manhattan Project's work was key to the development of nuclear power. Compton's work with the General Electric Company resulted in the development of the first fluorescent light bulb in 1934. His invention was critical to later developments in energy-efficient lighting, particularly the compact fluorescent light bulb.

Georges Jean Marie Darrieus (1888–1979)

Georges Darrieus was a French aeronautical engineer, best-known for his contributions to wind energy technology. In 1931, he patented the Darrieus turbine, a vertical axis wind turbine with long, thin blades connected at the top and bottom of the axis. Due to its likeness to an eggbeater, the Darrieus turbine is also known as the eggbeater windmill. Although it is not as widely used as the horizontal axis design, Darrieus' contribution encouraged the development of wind energy.

Rudolph Diesel (1858–1913)

Rudolph Diesel was a German inventor and mechanical engineer who is most famous for his invention of the diesel engine. He was educated at Munich Polytechnic, where he graduated with a degree in refrigeration engineering in 1879. His work in refrigeration inspired him to develop an engine that was more efficient than the steam engine. This desire was driven by sociological concerns for small-scale artisans who could not afford the large steam engines that were driving the Industrial Revolution. He designed many engines, including a solar-powered air engine. After years of development, he succeeded in creating the first diesel internal combustion engine in 1897, which was then patented in 1898. He became an instant millionaire because the diesel engine was much more efficient than the steam engine. He originally conceived of the engine running on local fuels derived from vegetable oil. Although the engine has been predominantly run on diesel fuel derived from petroleum, rising oil prices and environmental concerns have led to developments in biodiesel, bringing the diesel engine full circle back to the original conception that Rudolph Diesel had in mind when he invented it.

Thomas Alva Edison (1847–1931)

Thomas Edison is one of the most famous inventors in American history. He is best-known for his contributions to mass lighting through his development of centralized electricity. The United States and other industrialized nations adopted his design for the transmission of electricity, as well as, subsequently, by other nations as they entered the electrical age. His design has had a powerful impact on the energy sources chosen for electricity generation around the world. Only coal and nuclear power are capable of fulfilling the energy needs of industrialized countries in a centralized electrical design, making it difficult for renewable energy adoption. Renewable energy is the most efficient when used for decentralized and localized power needs. The problem of integrating renewable energy into a centralized power design has inhibited its adoption by major energy users like the United States. The problem of efficient power transmission from renewable energy sources has often been cited as one of the major obstacles to its widespread use. Although Edison's design has kept the world on a hard energy path, he was known as a socially conscious inventor who stressed the need to explore renewable energy sources to replace oil and coal.

John Ericsson (1803–1889)

John Ericsson was a Swedish inventor and mechanical engineer. He is best-known for inventing the ship propeller and his improvements to steam engine designs, particularly for ships. He was never formally educated but gained experience through his service in the Swedish army in 1820. After encountering several problems over the years with his ship engine designs based on traditional energy sources, he became interested in solar power. In the last 20 years of his life, he devoted his work to capturing solar energy out of fear that coal supplies would eventually run out. He succeeded in developing the first solar parabolic trough design. By 1888, he was on the verge of mass-producing an efficient solar engine for the purposes of irrigation. Unfortunately he died before he succeeded in mass-producing his solar motor, and the details of his design went with him because he never allowed for records of his designs to be kept until they were patented.

Michael Faraday (1791–1867)

Michael Faraday was a well-known English chemist and physicist who made many contributions to the fields of electromagnetism and electrochemistry. With little formal education, he worked his way up the ranks of scientific achievement and was eventually granted lifetime tenure as a Fullerian professor of chemistry in the Royal Society, a prestigious institute of scientific investigation in England. In 1833, he successfully explained what happened during electrolysis. His explanations became known as Faraday's laws and have formed the foundation of the modern understanding of electrolysis, making him known as the father of electrolysis. Electrolysis is the phenomenon by which electricity creates hydrogen, making his discoveries instrumental in the development of hydrogen as a carrier for renewable energy.

Enrico Fermi (1901–1954)

Enrico Fermi was an Italian physicist and an instrumental member of the Manhattan Project. Fermi received a doctorate in physics from the University of Pisa in 1922. He worked in various universities in Europe until earning a Nobel Prize for his work in 1938. Promptly after winning the Nobel Prize, he left his post at the University of Rome and went to the United States to escape Mussolini's fascist dictatorship. Once in the United States, he was appointed professor of physics at Columbia University. His work contributed to the discovery of nuclear fission, making him one of the greatest experts on neutrons in his time. After Otto Hahn and Frit Strassman discovered nuclear fission in 1939, Fermi begin working with Leo Szilard on the Manhattan Project at the University of Chicago to achieve a nuclear chain reaction. His work on the Manhattan Project ultimately led to the development of the first nuclear bomb and nuclear power.

Benoit Fourneyron (1802–1867)

Benoit Fourneyron was a French engineer who invented the first water turbine in 1827. At the age of 14, he graduated at the top of his class from the New School of Mines in Saint Etienne, France, in 1816. Working in collaboration with many of his past instructors, he began searching for ways to make the ancient waterwheel design more efficient for the purpose of powering the Industrial Revolution, which was just getting underway. He ultimately developed the first water turbine, which was a more efficient horizontal version of the waterwheel. The horizontal design, which allowed the turbine to fully utilize the power of the water motion, won him an award from the French Society for the Encouragement of Industry because the 6-horsepower turbine was the first commercial hydraulic turbine. With the prize money from the award, Fourneyron went on to improve his design, ending up with a 60-horsepower turbine that functioned at 80 percent efficiency. The design was adopted throughout Europe and the United States, becoming an important source of power for the Industrial Revolution. In 1895, Fourneyron turbines were installed at Niagara Falls to generate electricity.

Charles Fritts (d. 1903)

Charles Fritts was an American inventor from New York. Drawing on his experience as an electrician, Fritts built the first solar cell in 1883 out of selenium. He ambitiously believed that his photovoltaic cell would compete with Thomas Edison's new coal-fired electricity-generating plants. Unfortunately, his solar cells were able to convert solar power to electricity at an efficiency rate of less than 1 percent. Despite its inefficiency, Fritts's invention proved that it was possible to use sunlight to create electricity. Despite the fact the invention never competed with Edison's electric power plants, Fritts's work laid the necessary foundation for future developments in solar power technology.

Arnold J. Goldman (b. 1942)

Arnold J. Goldman is an American engineer and founder of Luz International Ltd. He received a degree in engineering from the

University of California, Los Angeles, and a master's degree in electrical engineering from the University of Southern California. In 1984, Goldman started Luz International Ltd., which designed, constructed, financed, and operated nine solar energy–generating stations in the Mojave Desert in California. By 1990 when the ninth plant was built, the company had succeeded in creating an economy of scale that resulted in a 50 percent reduction in cost per kilowatt of the ninth plant. The plants consisted of thousands of solar parabolic troughs that have generated over 11,000 gigawatt-hours and $1.7 billion in revenue since 1984, making them the first profitable large-scale solar projects. In 1991, Luz International filed for bankruptcy when it failed to obtain financing for the tenth solar thermal plant project. The bankruptcy was also attributed to the price drops in the oil and natural gas sectors and to repeals of government tax breaks for solar energy projects. Despite the bankruptcy, the nine plants continued operation. In 2004, Goldman established the company BrightSource Energy and is negotiating contracts in California to begin constructing solar projects using the solar power tower design. Lutz II, a Jerusalem-based subsidiary company of BrightSource, is building a power tower pilot project in the Negev Desert in Israel. Goldman's companies have succeeded in proving that large-scale solar energy projects can be profitable when government policies are in place to support development.

Albert A. Gore Jr. (b. 1948)

Albert A. Gore, known to the world as Al Gore, was born in Washington, D.C., to Albert Arnold Gore Sr., a U.S. representative (1939–1944, 1945–1953) and senator (1953–1971) from Tennessee, and Pauline LaFon Gore, one of the first women to graduate from Vanderbilt University Law School. Gore graduated from Harvard in 1969 with a BA in government. While at Harvard, Gore attended classes taught by Roger Revelle, the first scientist to link fossil fuel emissions to global warming. Revelle had a significant impact on Gore's views of the environment and has been cited by Gore as a major influence in his passion for the global warming issue. After serving in the U.S. military, Gore returned from Vietnam to attend Vanderbilt University but left without a degree and ran for Congress in Tennessee in 1976. He was reelected three more times and then ran a successful campaign for the U.S. Senate in 1984. During his tenure as a U.S. representative, Gore held the first congressional hearings on climate change and pushed the U.S. government to take measures to study and address its causes and effects. While serving in the Senate for nine consecutive years, Gore worked on his book, *Earth in the Balance: Ecology and the Human Spirit,* which was published in 1992 and provides policy recommendations to address global ecological challenges. The book was one of only a few written by a politician to make it onto *The New York Times* best-seller list.

Gore was elected vice president of the United States in 1992 and served two terms with President Bill Clinton. While serving as the vice president, Gore strongly pushed for the passage of the Kyoto Protocol. In 2000, Gore lost a bid for president to George W. Bush in a controversial election that was ultimately decided by the U.S. Supreme Court. After the defeat, Gore devoted much of his time and energy to raising awareness of the seriousness of climate change and implementing solutions in the private sector. In 2004, he started Generation Investment Management, a firm that assists firms already dedicated to preventing climate change with investment strategies that focus on a variety of solutions. In 2006, Gore's documentary, *An Inconvenient Truth,* was released. The documentary sheds light on human-caused global warming and argues for immediate action on the part of governments and individuals to address the problem. The documentary was awarded an Academy Award for best documentary and was one of the top grossing documentaries of all time. In 2007, Gore and the scientists working for the Intergovernmental Panel on Climate Change were awarded a joint Nobel Peace Prize for their efforts. Gore is internationally known as a renewable energy advocate and champion for environmental causes.

William R. Grove (1811–1896)

Sir William R. Grove was a British scientist known as a pioneer in fuel cell technology, although he pursued a career in law. He was born in Swansea, Wales, and studied classics at Brasenose College, Oxford. He received his degree in 1832 and began practicing law in 1835. Grove was married in 1837 and took a lengthy year-long honeymoon, during which he was able to pursue his passion—science—and wrote his first scientific paper on designing fuel cells. In 1839, he built a fuel cell, known as the Grove cell, and drew the attention of Michael Faraday who was, at the time, studying the phenomenon of electrolysis. Faraday invited Grove to present his findings to the prestigious Royal Institute, where he was subsequently offered a fellowship as the first professor of experimental philosophy. In 1842, he gave his inaugural lecture where he was the first to propose the theory of the conservation of energy, although he called it the *correlation of physical forces*. During his tenure, he built the first practical fuel cell, which provided electrical energy by combining hydrogen and oxygen, a key foundation for the development of modern-day fuel cells. Grove, like many scientists of his time, also suggested that there were energy sources beyond the dominant sources of his time and that had yet to be discovered, foreshadowing the renewable energy revolution of the late 19th century.

Subhendu Guha (b. 1941)

Subhendu Guha is an Indian American scientist known for his pioneering work in amorphous silicon and nanocrystalline silicon, which are used in solar energy applications. Born in Calcutta, India, Guha attended the Presidency College where he earned a degree in physics. He then went on to earn a PhD in electronics from the University of Calcutta in 1968, after which he went to work for the India-based Tata Institute of Fundamental Research (TIFR). At TIFR, his work on semiconductors led him to be interested in research on solar power. He became interested in the use of amorphous silicon as a material for photovoltaic cells because it is made from sand and does not have the toxic properties of other materials used in solar technology. After doing postdoctoral work in 1974 and 1975 at the University of Sheffield in England, he went back to TIFR and began researching a new way to produce amorphous silicon. He published his research findings in 1981, causing a major breakthrough in the solar energy technology field. His revolutionary technique for producing amorphous silicon is now used by most producers of the material. Afterward, he became interested in the practical applications of his research and took a position at a U.S.-based company called Energy Conversion Devices. In 1994, he combined his academic research experience on amorphous silicon with his industrial research on solar photovoltaics, and, by 1997, flexible solar roof shingles were on the market. Flexible solar shingles can be used as a roofing material and reflect state-of-the-art technology for converting sunlight to electricity on buildings. Although the cells are still not quite cost competitive in areas where electricity is produced through traditional means, they offer a more cost-effective way of bringing electricity to areas, especially in developing countries, that have yet to be electrified. They are already being used in Egypt, Mexico, and other developing countries around the world.

John Burden Sanderson Haldane (1892–1964)

John Burden Sanderson Haldane was a British geneticist and evolutionary biologist best-known as one of the founders of population genetics. Haldane graduated from the New College, Oxford, in 1911. After serving in World War I, Haldane went back to New College, as a fellow and then on to Cambridge University. After teaching at Cambridge for nine years, he accepted a professorship at the University College London. In the late 1950s, he moved to India, where he became a guru to the country's burgeoning scientific community. Haldane was an unusual and a distinctive man, qualities that inspired characters based on him in books by Aldous Huxley and C. S. Lewis. He was a popularizer of science and had an uncanny knack for predicting scientific advances. His greatest contribution to the field of renewable energy was a lecture he gave while at Cambridge University in 1923. In the lecture he suggested that liquefied hydrogen, produced from wind power by electrolysis, would be the energy carrier of the future. He was the first person to suggest this and spurred research in and development of hydrogen and renewable energy sources by other scientists.

James E. Hansen (b. 1941)

James E. Hansen is an American scientist who is best-known for his work in climatology. Hansen studied at the University of Iowa in the space science program under Dr. James Van Allen. He graduated with a degree in physics and mathematics in 1963, receiving the highest distinction. He remained at the University of Iowa and received a master's degree in astronomy in 1965 and a PhD in physics in 1967. Shortly after receiving his doctorate, he began conducting research on human-induced climate change, making him one of the first experts in the field. In 1981, Hansen published an article in *Science* magazine detailing his research on global warming and predicting that the effects of climate change would be evident as early as 1990. This was much sooner than the predictions of other climate scientists, who had projected that the effects would begin to be felt in 2020. Also in the paper, he made a correct prediction that it would be difficult to get politicians and the public to take action. After his paper was published, he began testifying about climate change before congressional committees, helping to raise more widespread awareness of the issue of global warming. Many members of the scientific community ridiculed Hansen for being a doomsayer, but they have since retracted their critiques as his predictions have come true. Hansen is one of the few climate scientists to argue that carbon dioxide has had less of an impact on warming than other greenhouse gases like methane, chlorofluorocarbons, and nitrous oxide. Despite this, his position is that emissions from fossil fuels must be reduced before carbon dioxide emissions tip the atmospheric balance even further. Hansen is well-known for his argument that the tipping point for irreversible climate change will occur in 2016 if drastic measures are not taken to bring greenhouse gas emissions under control. Although a criticizer of the Bill Clinton and George W. Bush administrations' positions on climate change, Hansen is one of Al Gore's top advisors and was referred to numerous times in Gore's award-winning documentary on climate change, *An Inconvenient Truth.* Hansen currently heads the NASA Goddard Institute for Space Studies in New York City and holds a position as an adjunct professor at Columbia University in the Department of Earth and Environmental Sciences. In 2005 and 2006, Hansen received considerable media attention for his claims that NASA administrators were trying to censor and influence his public statements on climate change, although top officials deny it. Despite the denial, NASA and other government agencies were pressured by the media, Congress, and the public to revise their media policies.

William Heronemus (1920–2002)

William Heronemus was an American engineer known for his contributions to wind energy technology. Heronemus received his BS from the U.S. Naval Academy and his MS in marine engineering from the Massachusetts Institute of Technology. In 1941, he began his service in the U.S. Navy, retiring as a captain in 1965. After being employed by United Aircraft Corporate Systems Center from 1965 to 1967, he taught civil engineering at the University of Massachusetts until 1985. While at the University of Massachusetts, Heronemus started the alternative energy program, now the Renewable Energy Research Laboratory. He led the team of researchers responsible for the development of a wind turbine that would be the basis for U.S. wind turbines today. He was a pioneer of the wind farm concept, although his interest lay mainly in offshore wind development. His offshore wind projects of the 1970s are being used in Europe as a guide for offshore wind power projects. Heronemus testified before congressional hearings in the 1970s as a wind power expert and has been called the father of American wind power.

M. King Hubbert (1903–1989)

M. King Hubbert was an American geophysicist who is best-known for his accurate prediction of peak oil production in the United States. Hubbert received a BS in geology from the University of Chicago in 1926. He remained at the university and received his MS in mathematics in 1928 and his PhD in physics in 1937. While pursuing his doctorate, he gained considerable experience teaching geophysics at Columbia University until 1940 and working on geophysical problems for the Amerada Petroleum Corporation in Oklahoma, the Illinois State Geological Survey, and the United States Geological Survey during his summers. After receiving his doctorate, he took a position at Shell Oil Company as director of the research laboratory. While working for the Shell Oil Company, Hubbert published a paper, "Energy from Fossil Fuels" (*Science,* 1949), in which he argued that the fossil fuel era would be much shorter than many supposed, creating a need for investment in other energy sources. In 1964, he left Shell Oil Company to work for the U.S. Geological Survey until retiring in 1976, all the while teaching at both Stanford University and the University of California, Berkeley. In 1956, he presented his theory on oil reserves, which became known as Hubbert's curve, in which he stated that oil and gas production would resemble a bell curve over time. This was one of the first solid scientific arguments for the finiteness of oil and gas. He used his theory to predict that the United States would hit peak oil production between 1966 and 1971, an argument that came to be known as Hubbert's peak. In 1975, the National Academy of Sciences confirmed that Hubbert's prediction was correct. Then in 1976, Hubbert predicted that global oil production would reach its peak in 1995. His prediction is unconfirmed and remains controversial.

Jeremy Leggett (b. 1954)

Jeremy Leggett is a British geologist, who is well-known for his work as an environmental activist and renewable energy advocate. He graduated with a PhD in earth sciences from Oxford. After receiving his doctorate, he began an academic career at Imperial College, where he researched earth history and acted as a consultant for the oil industry. In the 1980s, he went on to work at the London School of Mines, where major oil companies, like British Petroleum and Shell, funded his research. After witnessing the destruction that oil exploration was causing, in 1989 he left the business and joined Greenpeace as its principal scientific adviser on global warming. Disappointed by the lack of results the environmental advocacy group was achieving, he left Greenpeace in 1996 and founded Solar Century, the largest solar solutions company in the United Kingdom, in 1997. He is still the chief executive of the company and also serves on the U.K. government's Renewables Advisory Board. He is also a director of the world's first private equity renewable energy fund, Bank Sarasin's New Energies Invest AG. His work has been credited for being instrumental in bringing about global awareness of the climate change issue, and he is seen as a leading advocate of renewable energy technology. He has argued for years that global oil production is near its peak, creating a dire need for investment in renewable energy resources. He has written several books on energy and climate change, including *The Carbon War: Global Warming and the End of the Oil Era; Half Gone: Oil, Gas, Hot Air and the Global Energy Crisis;* and *The Empty Tank: Oil, Gas, Hot Air and the Coming Global Financial Catastrophe.*

Amory Bloch Lovins (b. 1947)

Amory Lovins, a well-known American environmentalist and renewable energy advocate, has been called by *Newsweek* one of the most influential energy thinkers in the Western world. His proficiency in science in high school led to his admittance to Harvard in 1964, but he left to study experimental physics at Magdalen College, Oxford, England. He then studied as a junior research fellow at Merton College, Oxford, where he earned his master's degree. He did not pursue a doctorate there because Oxford would not allow him to study energy. Since then, he has been awarded nine honorary doctorate degrees from distinguished universities around the world. In 1976, he received notoriety as an advocate for energy efficiency and renewable energy in an essay written for the journal *Foreign Affairs* titled "Energy Strategy: The Road Not Taken?" He has served as an adviser to the energy industry and the U.S. departments of Energy and Defense for over 30 years. Since 1990, he has led the development of ultra efficient automobiles and a profitable hydrogen transition strategy. Lovins has authored and coauthored several books on renewable energy and energy efficiency, including *Winning the Oil Endgame.*

He is currently the chair and chief scientist at the Rocky Mountain Institute, a nonprofit organization devoted to research on energy policy that he started with his former wife, Hunter Lovins, in 1982.

Auguste Mouchout (Mid-19th Century)

Auguste Mouchout was a 19th-century French inventor who taught mathematics at the Lyce de Tours. He was concerned with finding alternative sources of energy because of his country's dependence on coal. This interest led him to begin work in solar energy in 1860. His initial experiments involved concentrating solar power onto a cauldron that was encased in glass in order to produce steam. He improved the design by adding a reflector that concentrated the sun's rays, and he succeeded in using the apparatus to power a small steam engine in 1865. This made Mouchout the first known inventor to convert solar radiation into mechanical energy. He succeeded in impressing Emperor Napoleon III, who offered financial assistance for the development of an industrial-sized solar engine. Mouchout improved on his design by refining the reflector so that the solar energy could be even more concentrated and by adding a tracking device that enabled his machine to follow the sun. The improved design was well received by the French government, which thought it could be most useful in the tropical climate of the French colony Algeria. Mouchout was sent to Algeria, where he worked to enlarge his design. He then presented the new design back in Paris in 1878, and this time connected it to a refrigeration device, demonstrating that he could make ice from the sun's energy. Although this won him a medal, his funding was cut in 1880 due to the low price of coal and an improved transportation system. His groundbreaking research provided the foundation for the modern understanding of converting sunlight to mechanical energy.

Willie Nelson (b. 1933)

Willie Nelson is an American country singer, who is best-known for his style of outlaw country music developed in the 1970s. He is also a well-known environmental and biofuel advocate. In 2004, he partnered with his wife, Annie, and with Bob and Kelly King to build two Pacific Bio-diesel plants in Salem, Oregon, and Carl's Corner, Texas. He then founded Willie Nelson's Bio-diesel in 2005 with several other businesspeople. The company markets biodiesel biofuel to truck stops across the country. The fuel, which is called BioWillie, is made from vegetable oil derived from soybeans and can be used without modification in diesel engines. Efforts to encourage the use of BioWillie also coincide with Nelson's support of small American farmers, whose plight he has brought awareness to, since 1985, with an annual concert called Farm Aid. BioWillie thus serves three purposes: to help end U.S. dependence on foreign oil, to support American family farmers, and to improve the environment.

Roger Revelle (1909–1991)

Roger Revelle was an American scientist and one of the first to study global warming. He graduated from Pomona College in 1929 with a degree in geology and then received a PhD in oceanography at the University of California, Berkley. Revelle founded the Committee on Climate Change and the Ocean under the Scientific Committee on Ocean Research and International Ocean Commission. This led to the creation of the Atmospheric Carbon Dioxide Program, which started monitoring carbon dioxide in Hawaii and

Antarctica in 1956. In 1957, Revelle coauthored a paper with Hans Suess arguing that the oceans would be unable to absorb carbon dioxide at a rate fast enough to prevent it from having an effect on the atmosphere, meaning that there would be a greenhouse effect if large amounts of carbon dioxide are emitted into the atmosphere. The ocean's inability to absorb carbon dioxide became known as the Revelle factor. This was one of the first scientific studies to suggest that fossil fuel–induced carbon dioxide emissions could have a major effect on the climate. In the early 1980s, Revelle taught some of the first classes in academia that addressed the global warming issue. Revelle also taught these classes at Harvard University, where he inspired Al Gore's interest and concern about global warming.

Frank Shuman (1862–1918)

Frank Shuman was an American inventor responsible for developing the most cost-effective solar technology prior to the space age. Although he had very little formal education, Shuman's passion for scientific research landed him a position as a chemist at the Victor G. Bloede Company of West Virginia when he was just 18. In 1897, after years of development, Shuman developed a primitive parabolic trough that heated a chemical with a lower boiling point than water to produce steam and run an engine. In 1908, after making improvements to his design, Shuman established the Sun Power Company with the intention of producing larger versions of his model. His goal was to create a solar power plant that was capable of running industrial-sized steam engines. In 1911, he constructed the largest solar conversion system ever built near his home in Tacony, Pennsylvania. Although the structure was able to collect over 10,000 square feet of sunlight, it was still not sufficient to run the larger steam engines. After several improvements and collaborations with another engineer, E. P. Haines, Shuman was able to produce enough energy to power industrial applications. Like most renewable energy systems, the initial cost was much higher than that of fossil fuel sources, though in the long term, it was a much cheaper source because the fuel was free. Unfortunately, the short-term costs were the determining factor in Shuman's inability to get his invention adopted on a mass scale in North America. So he made plans to export his solar power plant to Africa. A machine more akin to modern-day parabolic trough designs was built in Cairo in 1912. The new machine was even more efficient than his previous efforts, and there was great potential for solar power to gain status as one of the dominant power sources. Unfortunately, Shuman's solar system was destroyed shortly afterward when World War I swept through the region. Fortunately, Shuman's design survived and was the foundation for successful modern-day solar parabolic troughs.

Lyman Spitzer Jr. (1914–1997)

Lyman Spitzer Jr. was an American astrophysicist who has been called one of the 20th century's greatest scientists. He is most famous for his contributions to space astronomy and was instrumental in the development of the Hubble Space Telescope. In the field of alternative energy, Spitzer was one of the first to study thermonuclear fusion. Spitzer received a degree in physics from Yale University in 1935, after which he spent a year at Cambridge University. In 1937, he earned a master's degree from Princeton University and a doctorate in astrophysics in 1938. He then spent a year as a postdoctorate fellow at Harvard University and went on to join the faculty at Yale in 1939. In 1951, after working on experiments to create a controlled nuclear fusion reaction, Spitzer founded the Princeton Plasma Physics Laboratory, originally called Project Matterhorn. This project was Princeton's pioneering nuclear fusion program and funded by the U.S. Atomic Energy Commission. Spitzer hoped to create a clean source of energy from fusion without the nuclear waste that fission reactions produced. Spitzer's research led to the development of the Tokamak Fusion Test Reactor, which made critical advances in scientific knowledge about fusion that were necessary for the International Thermonuclear Experimental Reactor (ITER), an international fusion test facility that will be completed in 2016 in France. ITER project leaders hope that after the 30-year experimental phase, nuclear fusion will be capable of providing for our energy needs.

Robert Stirling (1790–1878)

Robert Stirling was a Scottish clergyman and inventor of the Stirling engine, which is used in solar dish designs. Although a minister by profession, Stirling had a profound interest in engineering. His inventions were driven by a concern for his parishioners and the dangers many of them faced working around steam engines, which were known to explode. Stirling worked on an improved engine design in his home workshop in his spare time. In 1817, Stirling's *economiser* was patented by the Chancery in Edinburgh, Scotland. The design could not explode and worked at a lower pressure that prevented burns from steam. The first practical model was built in 1818 and became known as the Stirling engine. The engine is the highest-efficiency heat engine in existence even today, although it is not as cost-effective as internal combustion engines. Stirling never produced his engine on a mass scale because it was unable to compete with steam engines due to its costs, but the original design is used in contemporary solar applications. Most notably, Stirling Engine Systems, a California-based solar technology firm, has developed the Solar Stirling Dish System, which uses a solar dish to concentrate power onto a Stirling engine and produce electricity.

Alexander T. B. Stuart (1888–1950)

Alexander T. B. Stuart was a Canadian engineer and entrepreneur who conducted pioneering work in the development of hydrogen-based electrolysis. Stuart studied chemistry and mineralogy at the University of Toronto under Lash Miller, who had studied under William Grove, the inventor of the fuel cell. In 1905, Stuart began to take an interest in hydrogen as an energy source while working at the Niagara Falls Power Station. At the station, Stuart began to analyze the opportunity for a hydrogen energy economy, where he noted that only 30 to 40 percent of the capacity of the hydroelectric system was being used for lack of a way to store excess energy. He concluded that the excess energy should be used in a water electrolysis process to produce hydrogen fuel, which could then be used in a variety of energy applications. He spent the remainder of his life developing and commercializing energy technology. In 1920, he invented a water electrolysis device, the Stuart cell, which is still popular for use in hydrogen energy applications. In 1948, he formed Stuart Energy with his son, Alexander K. Stuart, to research and commercialize hydrogen energy applications. The company is still in business, run by his grandson Andrew T. B. Stuart, and is one of the largest and most successful hydrogen companies in the world.

Hans Eduard Suess (1909–1993)

Hans Eduard Suess was an Austrian chemist and physicist who contributed to the development of nuclear energy and the science of global warming. Suess earned his PhD in chemistry from the University of Vienna in 1935. During World War II, he was part of a team of German scientists studying atomic energy. He also advised the development of Norsk Hydro, a heavy water plant in Norway that was producing hydrogen by electrolyzing water. He emigrated to the United States in 1950 to continue his work in atomic energy. After coming to the United States, Suess became interested in carbon dating and began using it to study the climatic history of the Earth. In 1957, Roger Revelle, another pioneering climate change researcher who was interested in Suess's carbon dating research, invited Suess to join the faculty at the University of California, San Diego. That same year he coauthored a paper with Revelle arguing that the oceans would be unable to absorb carbon dioxide at a rate fast enough to prevent it from having an effect on the atmosphere, meaning that there would be a greenhouse effect if large amounts of carbon dioxide are emitted into the atmosphere. It was one of the first scientific papers to suggest that fossil fuel–induced carbon dioxide emissions could have a major effect on the climate. He also used his carbon dating method to determine the amount of dilution of carbon dioxide that had occurred in the atmosphere since the Industrial Revolution. The dilution is known as the Suess effect and has been used extensively in climate change measurements and studies for the past couple of decades.

Leo Szilard (1898–1964)

Leo Szilard was a Hungarian American physicist who discovered the nuclear chain reaction necessary for nuclear fission. He began his studies in engineering at the Budapest Technical University in 1916 but later transferred to the Berlin Institute of Technology, where he began studying physics. There he studied under Albert

Einstein, who thought that Szilard was one of his most promising students. In 1922, he received his doctorate in physics from Humboldt University in Berlin and was awarded the highest honor for his dissertation on thermodynamics. In 1933, Szilard conceived of the idea of a nuclear chain reaction while stopped at a traffic light, and in 1934 he filed to patent the idea. In 1938, he went to Columbia University in New York to conduct research and was soon joined by Enrico Fermi. Szilard and Fermi were recruited to work on the Manhattan Project at the University of Chicago, where they achieved a sustained nuclear chain reaction on December 2, 1942. The discovery led to the development of nuclear power. Szilard and Fermi are coholders of the patent for the nuclear fission reactor.

Tom Udall (b. 1948)

Tom Udall has served as a U.S. representative for New Mexico's Third Congressional District since 1999 and has proposed numerous pieces of renewable energy legislation throughout his tenure. Udall graduated with a prelaw degree in 1970 from Prescott College. He then attended Cambridge University, England, where he graduated in 1975 with a bachelor of law degree. He enrolled in the University of New Mexico School of Law, where he earned his juris doctor in 1977. Udall went on to serve as law clerk to Chief Justice Oliver Seth of the U.S. 10th Circuit Court of Appeals, an assistant U.S. attorney, chief counsel to the Department of Health and the Environment, and attorney general of New Mexico. He successfully ran for election in 1999 to the U.S. House of Representatives as a Democrat. Udall has sponsored several bills to establish a national renewable energy portfolio standard. Most recently, Udall successfully proposed an amendment to a house energy bill requiring electricity providers to produce no less than 15 percent of their electricity from renewable energy sources by 2020. The amendment and bill were passed by the House but not supported by the Senate. Udall is also a member of the United States House Peak Oil Caucus, which was formed in 2005 to address the need to make a transition to a sustainable energy economy.

6

Data and Documents

This chapter provides recent data on energy in general and on renewable energy use throughout the world and in the United States. All data has been extracted from the *Annual Energy Review 2006*, which is updated annually to include information on world energy use from 1946 or 1949 to the present. First, Table 6.1 presents the geographical and organizational definitions needed for any inquiry into the state of world affairs. An energy conversion table acts as a point of reference for comparing various estimates of energy use (Table 6.2). Following that, the numbers for total world energy consumption, world energy consumption by fuel, and world energy consumption by country are provided (Tables 6.3 through 6.12). This data reflects the broad picture of energy use and helps to identify individual characteristics of fuels and countries. The same is done for the United States by providing a look at overall U.S. renewable energy use and then at energy consumption by individual states (Tables 6.13 and 6.14). Concluding this chapter is data on global carbon emissions (Table 6.15).

Worldwide Data

TABLE 6.1
Geographical and Organizational Definitions

Central and South America

Antarctica	Dominican Republic	Nicaragua
Antigua and Barbuda	Ecuador	Panama
Argentina	El Salvador	Paraguay
Aruba	Falkland Islands	Peru
Bahamas, The	French Guyana	Puerto Rico
Barbados	Grenada	Saint Kitts and Nevis
Belize	Guadeloupe	Saint Lucia
Bolivia	Guatemala	Saint Vincent/Grenadines
Brazil	Guyana	Suriname
Cayman Islands	Haiti	Trinidad and Tobago
Chile	Honduras	Turks and Caicos Islands
Colombia	Jamaica	Uruguay
Costa Rica	Martinique	Venezuela
Cuba	Montserrat	Virgin Islands, British
Dominica	Netherlands Antilles	Virgin Islands, U.S.

North America

Bermuda	Greenland	Saint Pierre and Miquelon
Canada	Mexico	United States

Europe

Albania	France	Netherlands
Austria	Germany	Norway
Belgium	Germany, East	Poland
Bosnia and Herzegovina	Germany, West	Portugal
Bulgaria	Gibraltar	Romania
Croatia	Greece	Serbia

continues

TABLE 6.1 Continued
Geographical and Organizational Definitions

Europe (continued)		
Cyprus	Hungary	Slovakia
Czech Republic	Iceland	Slovenia
Denmark	Ireland	Spain
Faroe Islands	Italy	Sweden
Finland	Luxembourg	Switzerland
Former Czechoslovakia	Macedonia	Turkey
Former Serbia and Montenegro	Malta	United Kingdom
Former Yugoslavia	Montenegro	

Eurasia		
Armenia	Kazakhstan	Russia
Azerbaijan	Kyrgyzstan	Tajikistan
Belarus	Latvia	Turkmenistan
Estonia	Lithuania	Ukraine
Former USSR	Moldova	Uzbekistan
Georgia		

Middle East		
Bahrain	Kuwait	Saudi Arabia
Iran	Lebanon	Syria
Iraq	Oman	United Arab Emirates
Israel	Qatar	Yemen
Jordan		

Africa		
Algeria	Gabon	Reunion
Angola	Gambia, The	Rwanda
Benin	Ghana	Saint Helena
Botswana	Guinea	Sao Tome and Principe

continues

TABLE 6.1 Continued
Geographical and Organizational Definitions

Africa (continued)

Burkina Faso	Guinea-Bissau	Senegal
Burundi	Kenya	Seychelles
Cameroon	Lesotho	Sierra Leone
Cape Verde	Liberia	Somalia
Central African Republic	Libya	South Africa
Chad	Madagascar	Sudan
Comoros	Malawi	Swaziland
Congo (Brazzaville)	Mali	Tanzania
Congo (Kinshasa)	Mauritania	Togo
Cote d'Ivoire (Ivory Coast)	Mauritius	Tunisia
Djibouti	Morocco	Uganda
Egypt	Mozambique	Western Sahara
Equatorial Guinea	Namibia	Zambia
Eritrea	Niger	Zimbabwe
Ethiopia	Nigeria	

Asia and Oceania

Afghanistan	India	Niue
American Samoa	Indonesia	Pakistan
Australia	Japan	Papua New Guinea
Bangladesh	Kiribati	Philippines
Bhutan	Korea, North	Samoa
Brunei	Korea, South	Singapore
Burma	Laos	Solomon Islands
Cambodia	Macau	Sri Lanka
China	Malaysia	Taiwan
Cook Islands	Maldives	Thailand
East Timor	Mongolia	Tonga

continues

TABLE 6.1 Continued
Geographical and Organizational Definitions

Asia and Oceania (continued)		
Fiji	Nauru	U.S. Pacific Islands
French Polynesia	Nepal	Vanuatu
Guam	New Caledonia	Vietnam
Hawaiian Trade Zone	New Zealand	Wake Island
Hong Kong		

Organization for Economic Cooperation and Development (OECD)		
Australia	Guam	Norway
Austria	Hawaiian Trade Zone	Poland
Belgium	Hungary	Portugal
Canada	Iceland	Puerto Rico
Czech Republic	Ireland	Slovakia
Denmark	Italy	Spain
Finland	Japan	Sweden
Former Czechoslovakia	Korea, South	Switzerland
France	Luxembourg	Turkey
Germany	Mexico	United Kingdom
Germany, East	Netherlands	United States
Germany, West	New Zealand	Virgin Islands, U.S.
Greece		

OECD Europe		
Austria	Germany, West	Poland
Belgium	Greece	Portugal
Czech Republic	Hungary	Slovakia
Denmark	Iceland	Spain
Finland	Ireland	Sweden
Former Czechoslovakia	Italy	Switzerland

continues

TABLE 6.1 Continued
Geographical and Organizational Definitions

OECD Europe (continued)

France	Luxembourg	Turkey
Germany	Netherlands	United Kingdom
Germany, East	Norway	

International Energy Agency (IEA)

Australia	Greece	Norway
Austria	Guam	Portugal
Belgium	Hawaiian Trade Zone	Puerto Rico
Canada	Hungary	Spain
Czech Republic	Ireland	Sweden
Denmark	Italy	Switzerland
Finland	Japan	Turkey
France	Korea, South	United Kingdom
Germany	Luxembourg	United States
Germany, East	Netherlands	Virgin Islands, U.S.
Germany, West	New Zealand	

European Union (EU)

Austria	Germany	Malta
Belgium	Germany, East	Netherlands
Bulgaria	Germany, West	Poland
Cyprus	Greece	Portugal
Czech Republic	Hungary	Romania
Denmark	Ireland	Slovakia
Estonia	Italy	Slovenia
Finland	Latvia	Spain
Former Czechoslovakia	Lithuania	Sweden
France	Luxembourg	United Kingdom

continues

TABLE 6.1 Continued
Geographical and Organizational Definitions

Former USSR		
Armenia	Kazakhstan	Russia
Azerbaijan	Kyrgyzstan	Tajikistan
Belarus	Latvia	Turkmenistan
Estonia	Lithuania	Ukraine
Georgia	Moldova	Uzbekistan

Organization of Petroleum Exporting Countries (OPEC)		
Algeria	Iraq	Qatar
Angola	Kuwait	Saudi Arabia
Indonesia	Libya	United Arab Emirates
Iran	Nigeria	Venezuela

Source: Energy Information Administration, 2006.

TABLE 6.2
Energy Measurement Conversion Table

Energy Measurements	Joules	Btu	Watts	Kilowatts	Megawatts	Barrel of Oil
Joules	—	1055.0559	0.000277778	2.777777778^{-7}	2.777777778^{-10}	5305039787.8
Btu	1055.0559	—	0.293071083	0.00029307071	2.930710833^{-7}	5033241.3793
Watts	3600	3.41214148	—	0.001	0.000001	1473622.2812
Kilowatts	3,600,000	3412.141479897	1,000	—	0.001	1473.6222812
Megawatts	3,600,000,000	3,412,141.479896942	1,000,000	1000	—	1.4736222812

TABLE 6.3
Total World Energy Consumption
[Quadrillion (10^{15}) Btu]

Year	Energy Consumption
1980	283.4
1985	308.6
1990	347.3
1995	365.6
2000	399.6
2004	446.7
2010 (projected)	511.1
2015 (projected)	559.4
2020 (projected)	607.0
2025 (projected)	653.6
2030 (projected)	701.6

Source: Energy Information Administration.

TABLE 6.4
Total World Energy Consumption by Fuel Type:
Historical and Projected [Quadrillion (10^{15}) Btu]

Year	Oil	Natural Gas	Coal	Nuclear	Renewables
1980	131.0	54.0	70.2	7.6	18.4
1981	125.9	54.1	71.1	8.5	18.8
1982	122.7	54.2	72.5	9.5	19.3
1983	120.8	55.3	74.5	10.7	20.4
1984	123.3	61.0	78.7	13.0	21.0
1985	123.1	63.6	82.4	15.3	21.2
1986	127.2	64.3	83.6	16.2	21.8
1987	129.7	67.7	86.3	17.6	21.9
1988	133.7	71.1	89.1	19.2	22.5
1989	135.4	74.3	89.0	19.7	23.1
1990	136.2	75.2	89.4	20.4	24.0
1991	137.2	76.6	86.1	21.2	24.6
1992	137.9	76.8	85.5	21.3	24.7
1993	137.7	78.9	86.6	22.0	25.9
1994	140.2	78.8	87.4	22.4	26.2
1995	142.4	81.0	89.1	23.3	27.5
1996	145.7	84.6	90.1	24.1	28.1
1997	149.1	84.8	92.1	23.9	28.5
1998	150.5	85.8	90.4	24.3	28.7
1999	153.4	88.2	90.9	25.1	29.3
2000	155.4	91.4	94.9	25.7	30.0
2001	156.8	92.8	96.1	26.4	29.5
2002	158.1	96.1	97.0	26.7	29.9
2003	161.5	99.6	105.6	26.4	30.5
2004	167.7	103.8	114.5	27.5	33.7
2005	170.8	106.4	120.4	28.8	36.1
2006	173.3	108.5	123.7	29.1	37.2
2007	175.9	111.9	126.9	29.3	38.1
2008	178.8	115.0	130.0	29.5	38.9

continues

TABLE 6.4 Continued
Total World Energy Consumption by Fuel Type:
Historical and Projected [Quadrillion (10^{15}) Btu]

Year	Oil	Natural Gas	Coal	Nuclear	Renewables
2009	181.8	117.8	133.1	29.6	39.7
2010	183.9	120.6	136.3	29.8	40.5
2011	187.6	123.2	139.6	30.3	41.1
2012	190.4	126.1	142.7	30.9	41.7
2013	193.2	128.8	145.7	31.5	42.2
2014	195.8	131.5	148.8	32.0	42.8
2015	197.6	134.3	151.6	32.5	43.4
2016	201.2	136.9	154.8	33.2	44.1
2017	203.7	139.3	157.9	33.8	44.7
2018	206.4	141.7	161.0	34.5	45.3
2019	209.1	144.3	164.0	35.1	46.0
2020	210.6	147.0	167.2	35.7	46.6
2021	214.5	149.2	170.3	36.1	47.2
2022	217.2	151.5	173.5	36.6	47.9
2023	220.0	153.8	176.6	37.1	48.6
2024	222.7	156.1	179.8	37.6	49.4
2025	224.1	158.5	182.9	38.1	50.1
2026	228.3	160.8	186.3	38.4	50.8
2027	231.3	163.2	189.6	38.8	51.4
2028	234.2	165.8	192.8	39.1	52.1
2029	237.1	168.1	196.0	39.5	52.8
2030	238.9	170.4	199.1	39.7	53.5

Source: Energy Information Administration.

TABLE 6.5
World Consumption of Primary Energy by Energy Type and Selected Country Groups,
2000–2005

Energy Type/Country Group	2000	2001	2002	2003	2004	2005
Petroleum (thousand barrels per day)						
World total	**76,626**	**77,371**	**78,016**	**79,593**	**82,304**	**83,607**
OECD	47,840	47,916	47,870	48,586	49,331	49,617
Non-OECD	28,786	29,456	30,146	31,007	32,973	33,991
Other groups						
OECD Europe	15,159	15,341	15,284	15,424	15,467	15,515
OPEC	6,119	6,448	6,709	6,852	7,219	7,651
EU	14,540	14,764	14,686	14,808	14,867	14,925
IEA	45,308	45,413	45,416	46,112	46,782	46,978
Dry Natural Gas (trillion cubic feet)						
World total	**88.275**	**89.499**	**92.653**	**95.960**	**99.836**	**103.700**
OECD	48.683	48.305	49.775	50.519	51.617	51.966
Non-OECD	39.592	41.194	42.878	45.441	48.219	51.734
Other groups						
OECD Europe	16.437	16.908	17.161	18.166	18.925	19.291
OPEC	9.281	9.758	10.231	10.460	11.211	12.510
EU	16.685	17.213	17.306	18.165	18.813	19.110
IEA	46.560	46.158	47.473	48.049	49.119	49.414
Coal (million short tons)						
World total	**5,098**	**5,191**	**5,272**	**5,706**	**6,150**	**6,483**
OECD	2,462	2,445	2,459	2,514	2,542	2,568
Non-OECD	2,635	2,746	2,813	3,192	3,608	3,915
Other groups						
OECD Europe	913	908	896	911	909	905
OPEC	26	33	36	37	43	48
EU	910	919	912	939	933	909
IEA	2,279	2,267	2,283	2,329	2,362	2,388

continues

TABLE 6.5 Continued
World Consumption of Primary Energy by Energy Type and Selected Country Groups, 2000–2005

Net Hydroelectric Power (billion kilowatt-hours)

World total	**2,645.4**	**2,550.7**	**2,596.8**	**2,616.0**	**2,759.2**	**2,900.0**
OECD	1,330.2	1,233.7	1,252.3	1,226.4	1,254.5	1,258.5
Non-OECD	1,315.1	1,317.1	1,344.5	1,389.6	1,504.7	1,641.5

Other groups

OECD Europe	535.7	533.5	490.7	458.9	483.7	480.9
OPEC	83.0	84.1	86.8	89.5	99.8	111.5
EU	349.3	369.0	312.1	299.8	317.0	301.3
IEA	1,284.5	1,191.8	1,213.2	1,194.6	1,216.3	1,217.4

Net Nuclear Electric Power (billion kilowatt-hours)

World total	2,449.9	2,516.7	2,545.3	2,517.8	2,615.0	2,625.6
OECD	2,128.2	2,176.1	2,177.7	2,127.5	2,216.2	2,226.6
Non-OECD	321.7	340.6	367.6	390.2	398.8	399.0

Other groups

OECD Europe	887.9	915.8	923.1	931.5	940.5	929.0
OPEC	0	0	0	0	0	0
EU	897.8	929.4	941.3	945.6	955.4	944.9
IEA	2,104.7	2,151.6	2,151.3	2,100.6	2,191.3	2,199.4

Net Geothermal, Solar, Wind, and Wood and
Waste Electric Power (billion kilowatt-hours)

World total	**242.6**	**253.0**	**284.5**	**308.2**	**341.5**	**369.7**
OECD	199.1	205.5	234.0	254.8	284.1	309.6
Non-OECD	43.5	47.5	50.4	53.4	57.3	60.1

Other groups

OECD Europe	75.4	84.1	99.7	115.9	140.6	160.0
OPEC	4.6	5.7	5.9	6.0	6.3	6.3
EU	72.1	80.6	96.3	112.1	136.5	155.5
IEA	191.2	197.4	226.0	244.2	272.8	296.6

Source: Energy Information Administration.

TABLE 6.6
World Petroleum Consumption by Country, 2000–2005 (Thousand Barrels per Day)

Region/Country	2000	2001	2002	2003	2004	2005
North America						
Bermuda	3.66	3.63	3.67	3.98	4.25	4.40
Canada	2,026.67	2,056.84	2,078.35	2,207.15	2,301.75	2,289.87
Greenland	3.65	3.77	3.81	3.83	3.86	3.88
Mexico	2,036.36	2,008.69	1,949.76	1,948.58	1,995.92	2,078.23
Saint Pierre and Miquelon	0.50	0.46	0.46	0.50	0.54	0.55
United States	19,701.08	19,648.71	19,761.30	20,033.50	20,731.15	20,802.16
North American totals	**23,771.93**	**23,722.10**	**23,797.35**	**24,197.55**	**25,037.48**	**25,179.09**
Central and South America						
Antarctica	1.47	1.49	1.49	1.53	1.53	1.55
Antigua and Barbuda	3.50	3.51	3.53	3.74	4.04	4.00
Argentina	510.93	474.39	438.30	449.99	471.79	480.00
Aruba	6.51	6.53	6.53	6.98	7.06	7.00
Bahamas, The	22.22	21.84	22.26	26.60	26.96	26.00
Barbados	10.82	10.37	10.37	10.57	7.10	9.00
Belize	4.71	6.26	6.01	6.29	6.75	7.00
Bolivia	48.10	46.74	46.72	46.26	49.62	52.00
Brazil	2,166.28	2,206.08	2,131.60	2,055.68	2,122.81	2,166.00
Cayman Islands	2.39	2.41	2.41	2.50	2.70	2.70
Chile	235.90	232.49	233.96	235.77	244.63	250.00
Colombia	277.49	271.18	261.20	265.36	267.52	264.00
Costa Rica	36.28	37.59	39.66	41.92	41.56	43.00
Cuba	195.00	203.60	201.51	204.85	201.22	202.00
Dominica	0.59	0.75	0.82	0.88	0.77	0.80
Dominican Republic	110.06	113.95	116.15	115.28	114.07	116.00
Ecuador	130.54	139.97	141.73	146.06	151.90	155.00
El Salvador	38.06	39.41	39.01	42.36	41.93	43.20
Falkland Islands	0.19	0.19	0.21	0.21	0.23	0.24

continues

TABLE 6.6 Continued
World Petroleum Consumption by Country, 2000–2005 (Thousand Barrels per Day)

French Guiana	6.34	6.35	6.53	6.81	6.62	6.70
Grenada	0.96	1.61	1.74	1.76	1.78	1.80
Guadeloupe	12.43	12.66	12.88	13.39	13.72	14.00
Guatemala	59.29	65.90	64.56	64.98	67.57	70.00
Guyana	10.81	11.10	11.27	11.08	10.07	10.50
Haiti	10.33	11.23	11.60	11.55	11.84	12.00
Honduras	28.11	33.00	36.34	36.54	41.85	43.00
Jamaica	65.64	66.20	67.78	70.43	71.33	72.00
Martinique	13.35	13.46	13.67	14.24	14.67	15.10
Montserrat	0.37	0.37	0.37	0.42	0.46	0.48
Netherlands Antilles	70.96	71.86	70.51	65.76	66.39	68.00
Nicaragua	24.02	25.58	25.42	25.64	27.17	28.00
Panama	77.38	80.28	76.59	77.57	88.03	90.00
Paraguay	25.10	24.06	25.05	26.02	26.09	27.00
Peru	156.79	153.87	152.34	154.22	161.96	166.00
Puerto Rico	201.35	213.86	212.05	217.81	222.76	230.00
Saint Kitts and Nevis	0.70	0.70	0.70	0.79	0.87	0.90
Saint Lucia	2.33	2.44	2.49	2.74	2.68	2.70
Saint Vincent/Grenadines	1.21	1.23	1.31	1.38	1.47	1.50
Suriname	10.33	10.88	11.46	11.73	12.22	12.00
Trinidad and Tobago	24.68	29.00	28.31	31.34	32.45	28.00
Turks and Caicos Islands	NA	0.11	0.08	0.08	0.08	0.08
Uruguay	43.18	34.11	37.04	36.69	38.47	39.00
Venezuela	499.71	544.47	570.67	540.64	552.85	598.00
Virgin Islands, U.S.	66.08	91.70	93.24	108.78	110.90	98.00
Virgin Islands, British	0.41	0.41	0.41	0.48	0.60	0.60
Central and South American totals	**5,212.89**	**5,325.20**	**5,237.88**	**5,195.68**	**5,349.07**	**5,464.85**

continues

TABLE 6.6 Continued
World Petroleum Consumption by Country, 2000–2005 (Thousand Barrels per Day)

Europe						
Albania	21.04	22.68	24.28	27.40	28.83	29.00
Austria	245.60	262.08	270.33	288.03	286.83	295.13
Belgium	587.35	596.05	600.59	627.32	604.87	564.02
Bosnia and Herzegovina	19.23	19.28	20.13	21.77	24.94	26.00
Bulgaria	99.88	101.31	105.73	102.21	104.45	108.00
Croatia	85.54	85.54	89.49	91.34	95.50	99.00
Cyprus	47.50	51.63	50.88	51.60	54.54	56.00
Czech Republic	169.81	179.15	176.00	187.57	205.88	212.95
Denmark	210.00	213.41	197.16	188.27	185.33	183.46
Faroe Islands	4.46	4.46	4.49	4.53	4.58	4.60
Finland	205.25	206.64	217.16	222.35	221.58	219.75
Former Czechoslovakia	—	—	—	—	—	—
Former Serbia and Montenegro	62.48	81.94	83.82	84.67	85.08	85.00
Former Yugoslavia	—	—	—	—	—	—
France	2,000.51	2,051.77	1,982.83	1,999.04	2,006.60	1,999.17
Germany	2,771.85	2,814.62	2,721.64	2,678.72	2,665.48	2,618.05
Germany, East	—	—	—	—	—	—
Germany, West	—	—	—	—	—	—
Gibraltar	41.09	22.68	23.07	23.67	24.35	25.00
Greece	399.21	405.73	408.39	428.73	419.76	415.71
Hungary	143.25	138.16	140.39	135.22	138.22	152.23
Iceland	18.23	17.39	18.16	18.21	19.00	18.46
Ireland	170.20	182.62	180.45	175.83	182.07	192.02
Italy	1,853.77	1,836.85	1,870.13	1,873.27	1,793.87	1,731.97
Luxembourg	47.55	50.62	51.69	55.72	61.18	64.02
Macedonia	22.07	18.48	20.17	19.80	19.65	20.00
Malta	18.16	15.06	18.05	17.98	18.21	18.60
Montenegro	—	—	—	—	—	—
Netherlands	854.52	893.65	898.32	918.64	947.87	1,023.68

continues

TABLE 6.6 Continued
World Petroleum Consumption by Country, 2000–2005 (Thousand Barrels per Day)

Norway	210.36	220.10	217.21	232.04	219.95	228.38
Poland	411.26	404.69	406.48	431.92	459.08	462.75
Portugal	332.66	333.81	343.18	325.92	327.91	335.45
Romania	224.24	228.91	232.28	219.61	225.09	236.00
Serbia	—	—	—	—	—	—
Slovakia	66.77	71.67	80.07	74.94	75.10	79.35
Slovenia	52.28	52.02	51.04	52.00	52.62	54.00
Spain	1,433.20	1,492.35	1,504.53	1,542.38	1,573.18	1,599.89
Sweden	329.75	343.38	349.55	361.53	358.99	363.16
Switzerland	274.03	276.90	267.58	269.89	269.73	274.99
Turkey	666.88	618.62	657.73	644.97	661.37	660.85
United Kingdom	1,757.11	1,730.37	1,723.99	1,743.39	1,783.07	1,819.96
European totals	**15,857.07**	**16,044.61**	**16,007.00**	**16,140.46**	**16,204.74**	**16,276.58**
Eurasia						
Armenia	35.21	37.09	38.64	39.91	41.24	40.00
Azerbaijan	136.89	119.89	110.35	110.78	112.60	115.00
Belarus	139.77	130.33	123.44	142.83	158.46	156.00
Estonia	22.93	23.54	23.52	24.36	27.74	29.00
Former USSR	—	—	—	—	—	—
Georgia	15.84	12.67	12.29	12.44	12.97	13.40
Kazakhstan	194.75	210.45	217.16	206.98	221.25	234.00
Kyrgyzstan	10.57	9.25	10.09	9.57	12.23	12.00
Latvia	28.86	29.75	28.23	29.82	32.63	34.00
Lithuania	56.53	54.64	53.04	51.56	55.07	57.00
Moldova	9.58	10.56	11.98	12.91	14.30	14.50
Russia	2,578.50	2,590.23	2,636.41	2,681.86	2,750.81	2,757.00
Tajikistan	23.33	24.96	25.45	27.49	29.49	30.00
Turkmenistan	62.36	74.15	78.46	86.78	94.55	98.00
Ukraine	260.14	304.46	308.03	322.46	325.34	328.00
Uzbekistan	146.09	148.26	151.36	150.47	152.11	155.00
Eurasian totals	**3,721.35**	**3,780.21**	**3,828.45**	**3,910.22**	**4,040.80**	**4,072.90**

continues

TABLE 6.6 Continued
World Petroleum Consumption by Country, 2000–2005 (Thousand Barrels per Day)

Middle East

Bahrain	23.29	23.98	23.92	25.95	29.09	31.00
Iran	1,248.32	1,285.32	1,350.35	1,425.65	1,488.52	1,572.00
Iraq	462.32	489.08	497.37	457.38	502.66	545.00
Israel	255.36	273.36	246.46	251.99	241.35	236.00
Jordan	101.08	98.84	103.33	106.43	106.98	109.00
Kuwait	264.42	275.18	285.16	297.83	305.32	333.00
Lebanon	105.53	101.32	101.92	104.49	102.29	106.00
Oman	52.54	54.72	57.23	58.43	63.91	66.00
Qatar	48.17	54.48	62.25	71.88	82.85	95.00
Saudi Arabia	1,537.10	1,606.30	1,676.25	1,774.59	1,884.41	2,000.00
Syria	255.52	253.73	256.08	257.10	257.62	260.00
United Arab Emirates	330.48	324.65	334.63	333.01	351.14	372.00
Yemen	96.62	102.45	112.31	121.49	123.28	128.00
Middle Eastern totals	**4,780.75**	**4,943.41**	**5,107.25**	**5,286.23**	**5,539.41**	**5,853.00**

Africa

Algeria	206.18	217.91	224.56	228.77	238.35	250.00
Angola	29.26	43.04	44.89	47.46	48.29	50.00
Benin	11.37	11.40	12.13	14.16	16.77	16.00
Botswana	13.97	11.53	11.08	11.08	13.49	12.00
Burkina Faso	7.48	7.58	7.78	8.05	8.16	8.30
Burundi	2.64	2.80	2.83	2.94	2.69	2.90
Cameroon	22.73	22.54	22.45	23.50	23.52	24.20
Cape Verde	1.50	1.05	1.12	1.12	2.08	2.00
Central African Republic	2.20	2.24	2.33	2.33	2.20	2.30
Chad	1.34	1.37	1.41	1.41	1.32	1.35
Comoros	0.64	0.67	0.69	0.69	0.71	0.70
Congo (Brazzaville)	4.40	5.59	4.83	6.04	6.71	7.00
Congo (Kinshasa)	14.16	9.53	8.09	8.13	11.32	11.00
Cote d'Ivoire (Ivory Coast)	32.90	21.72	23.15	22.91	26.87	27.00

continues

TABLE 6.6 Continued

World Petroleum Consumption by Country, 2000–2005 (Thousand Barrels per Day)

Djibouti	11.21	11.38	11.40	11.65	11.84	12.00
Egypt	552.80	544.67	564.70	561.11	603.92	635.00
Equatorial Guinea	1.53	1.15	1.15	1.19	1.03	1.00
Eritrea	4.16	4.93	4.59	5.18	4.87	5.00
Ethiopia	22.73	23.31	25.64	27.55	28.59	29.00
Gabon	12.14	12.53	11.94	12.58	12.72	13.00
Gambia, The	1.79	1.92	1.97	1.99	2.01	2.03
Ghana	36.96	36.27	38.65	41.54	45.30	47.00
Guinea	8.21	8.30	8.36	8.40	8.48	8.50
Guinea-Bissau	2.30	2.32	2.41	2.47	2.46	2.48
Kenya	56.80	51.93	51.17	54.20	59.18	64.00
Lesotho	1.40	1.40	1.40	1.40	1.40	1.40
Liberia	2.94	3.10	3.36	3.46	3.51	3.55
Libya	210.28	225.17	233.35	244.09	254.51	266.00
Madagascar	12.13	13.32	15.31	15.84	17.39	17.00
Malawi	5.26	5.35	5.41	5.43	6.26	6.00
Mali	3.84	3.85	4.20	4.24	4.37	4.50
Mauritania	23.76	23.74	22.75	23.30	19.96	20.00
Mauritius	19.88	21.75	19.79	20.55	21.38	22.00
Morocco	158.48	161.22	164.28	167.07	170.29	176.00
Mozambique	7.81	9.37	10.65	11.19	13.32	13.00
Namibia	12.63	15.05	15.42	16.64	17.58	18.40
Niger	4.98	5.07	5.30	5.40	5.41	5.45
Nigeria	245.57	305.69	303.95	288.47	277.07	300.00
Reunion	17.72	18.53	18.49	18.50	18.82	19.00
Rwanda	5.19	5.24	5.27	5.33	5.16	5.30
Saint Helena	0.16	0.09	0.09	0.09	0.06	0.07
Sao Tome and Principe	0.61	0.62	0.64	0.64	0.63	0.65
Senegal	29.55	29.44	30.05	30.32	33.85	35.00
Seychelles	3.84	3.95	5.72	5.54	5.80	6.00
Sierra Leone	6.15	6.29	6.40	6.55	7.79	8.00

continues

TABLE 6.6 Continued
World Petroleum Consumption by Country, 2000–2005 (Thousand Barrels per Day)

Somalia	4.80	4.80	4.80	4.80	4.80	5.00
South Africa	457.93	458.24	475.44	490.21	503.91	497.00
Sudan	43.11	55.99	58.30	65.42	71.21	72.00
Swaziland	3.53	3.53	3.53	3.53	3.53	3.50
Tanzania	16.17	18.95	21.73	23.21	24.80	25.00
Togo	9.47	6.82	8.80	14.20	15.13	16.00
Tunisia	84.55	87.22	87.86	87.66	89.40	90.00
Uganda	8.56	9.52	9.93	10.69	10.87	11.00
Western Sahara	1.68	1.70	1.70	1.70	1.70	1.75
Zambia	10.74	11.76	12.19	12.74	13.27	14.00
Zimbabwe	25.35	23.46	22.32	20.41	13.37	16.00
African totals	**2,499.44**	**2,597.91**	**2,667.73**	**2,715.09**	**2,819.46**	**2,912.33**
Asia and Oceania						
Afghanistan	4.72	4.69	4.79	3.81	4.12	5.00
American Samoa	3.74	3.75	3.82	3.82	3.81	4.00
Australia	872.22	874.34	884.69	889.09	885.45	903.15
Bangladesh	68.80	81.25	82.77	83.87	84.64	86.00
Bhutan	0.98	1.05	1.09	1.13	1.14	1.20
Brunei	12.05	11.59	11.47	12.94	12.95	13.00
Burma	36.75	33.65	34.52	35.97	40.29	42.00
Cambodia	3.52	3.58	3.60	3.68	3.59	3.70
China	4,795.71	4,917.88	5,160.71	5,578.11	6,437.48	6,720.00
Cook Islands	0.39	0.39	0.39	0.39	0.43	0.45
East Timor	—	—	—	NA	NA	NA
Fiji	5.39	5.41	8.22	10.26	8.60	9.00
French Polynesia	4.58	4.59	4.82	5.73	5.68	5.80
Guam	18.58	20.21	13.66	15.40	12.13	13.53
Hawaiian Trade Zone	—	—	—	—	—	—
Hong Kong	244.92	245.38	272.63	276.58	318.15	293.00
India	2,127.44	2,183.73	2,263.44	2,346.33	2,429.62	2,438.00
Indonesia	1,036.70	1,077.00	1,125.65	1,142.67	1,232.57	1,270.00

continues

TABLE 6.6 Continued
World Petroleum Consumption by Country, 2000–2005 (Thousand Barrels per Day)

Japan	5,491.79	5,396.43	5,303.51	5,415.93	5,294.81	5,353.19
Kiribati	0.17	0.17	0.22	0.22	0.22	0.22
Korea, North	25.38	22.51	24.53	24.52	24.02	24.00
Korea, South	2,135.29	2,132.04	2,149.15	2,175.42	2,155.12	2,175.56
Laos	2.73	2.74	2.85	2.94	2.90	2.95
Macau	10.12	10.73	11.73	12.26	15.03	15.26
Malaysia	465.02	475.10	462.75	479.86	508.04	501.00
Maldives	3.14	3.37	6.19	6.96	4.87	5.00
Mongolia	8.50	9.86	10.67	10.35	12.28	12.00
Nauru	0.96	0.97	1.01	1.01	1.02	1.05
Nepal	14.90	15.26	15.51	15.67	16.15	16.60
New Caledonia	8.59	8.61	10.02	10.02	11.37	11.00
New Zealand	131.63	132.26	141.09	149.97	154.49	157.58
Niue	0.02	0.02	0.02	0.02	0.02	0.02
Pakistan	365.01	360.12	355.89	336.60	326.85	345.00
Papua New Guinea	14.94	14.98	22.04	25.18	25.02	26.00
Philippines	352.77	346.85	337.75	332.61	337.22	340.00
Samoa	0.97	0.97	1.02	1.02	1.06	1.10
Singapore	660.30	707.56	698.04	668.30	745.66	802.00
Solomon Islands	1.21	1.21	1.25	1.25	1.30	1.30
Sri Lanka	74.60	73.38	76.41	78.27	80.99	84.00
Taiwan	865.30	881.72	893.74	929.90	948.02	970.00
Thailand	724.94	701.61	763.27	832.32	915.47	929.00
Tonga	0.92	0.88	0.78	0.78	0.84	0.88
U.S. Pacific Islands	2.01	2.01	2.01	2.01	2.01	2.00
Vanuatu	0.56	0.57	0.61	0.61	0.63	0.64
Vietnam	175.71	178.57	192.90	214.62	238.37	254.00
Wake Island	8.98	9.00	9.16	9.16	9.13	9.30
Asian and Oceanian totals	**20,782.96**	**20,958.01**	**21,370.37**	**22,147.54**	**23,313.54**	**23,848.47**
World totals	**76,626.37**	**77,371.45**	**78,016.05**	**79,592.79**	**82,304.50**	**83,607.22**

Source: Energy Information Administration.

TABLE 6.7
World Natural Gas Consumption, 2000–2005 [Quadrillion (10^{15}) Btu]

Region/Country	2000	2001	2002	2003	2004	2005
North America						
Bermuda	0	0	0	0	0	0
Canada	3.051	3.190	3.243	3.457	3.424	3.501
Greenland	0	0	0	0	0	0
Mexico	1.471	1.492	1.679	1.818	1.831	1.870
Saint Pierre and Miquelon	0	0	0	0	0	0
United States	23.916	22.862	23.628	22.968	22.994	22.886
North American totals	**28.438**	**27.543**	**28.550**	**28.243**	**28.249**	**28.257**
Central and South America						
Antarctica	0	0	0	0	0	0
Antigua and Barbuda	0	0	0	0	0	0
Argentina	1.226	1.152	1.117	1.276	1.399	1.492
Aruba	0	0	0	0	0	0
Bahamas, The	0	0	0	0	0	0
Barbados	0.001	0.001	0.001	0.001	0.001	0.001
Belize	0	0	0	0	0	0
Bolivia	0.046	0.030	0.038	0.083	0.079	0.078
Brazil	0.347	0.412	0.492	0.519	0.632	0.684
Cayman Islands	0	0	0	0	0	0
Chile	0.193	0.235	0.239	0.293	0.307	0.317
Colombia	0.187	0.200	0.202	0.200	0.203	0.219
Costa Rica	0	0	0	0	0	0
Cuba	0.022	0.011	0.013	0.014	0.014	0.015
Dominica	0	0	0	0	0	0
Dominican Republic	0	0	0	0.011	0.005	0.009
Ecuador	0.006	0.007	0.005	0.007	0.008	0.012
El Salvador	0	0	0	0	0	0
Falkland Islands	0	0	0	0	0	0

continues

TABLE 6.7 Continued
World Natural Gas Consumption, 2000–2005 [Quadrillion (10¹⁵) Btu]

French Guiana	0	0	0	0	0	0
Grenada	0	0	0	0	0	0
Guadeloupe	0	0	0	0	0	0
Guatemala	0	0	0	0	0	0
Guyana	0	0	0	0	0	0
Haiti	0	0	0	0	0	0
Honduras	0	0	0	0	0	0
Jamaica	0	0	0	0	0	0
Martinique	0	0	0	0	0	0
Montserrat	0	0	0	0	0	0
Netherlands Antilles	0	0	0	0	0	0
Nicaragua	0	0	0	0	0	0
Panama	0	0	0	0	0	0
Paraguay	0	0	0	0	0	0
Peru	0.011	0.012	0.014	0.017	0.028	0.052
Puerto Rico	0.013	0.023	0.023	0.027	0.025	0.025
Saint Kitts and Nevis	0	0	0	0	0	0
Saint Lucia	0	0	0	0	0	0
Saint Vincent/Grenadines	0	0	0	0	0	0
Suriname	0	0	0	0	0	0
Trinidad and Tobago	0.370	0.415	0.442	0.473	0.522	0.600
Turks and Caicos Islands	0	0	0	0	0	0
Uruguay	0.001	0.001	0.001	0.002	0.004	0.003
Venezuela	1.144	1.334	1.195	1.026	1.144	1.207
Virgin Islands, U.S.	0	0	0	0	0	0
Virgin Islands, British	0	0	0	0	0	0
Central and South American totals	**3.568**	**3.834**	**3.782**	**3.950**	**4.372**	**4.714**
Europe						
Albania	0.001	0.001	0.001	0.001	0.001	0.001
Austria	0.288	0.305	0.308	0.333	0.336	0.360

continues

TABLE 6.7 Continued
World Natural Gas Consumption, 2000–2005 [Quadrillion (10¹⁵) Btu]

Belgium	0.590	0.582	0.600	0.639	0.652	0.651
Bosnia and Herzegovina	0.011	0.011	0.006	0.006	0.011	0.014
Bulgaria	0.192	0.204	0.173	0.190	0.187	0.190
Croatia	0.102	0.105	0.097	0.113	0.102	0.099
Cyprus	0	0	0	0	0	0
Czech Republic	0.330	0.354	0.342	0.346	0.343	0.339
Denmark	0.206	0.204	0.204	0.206	0.206	0.197
Faroe Islands	0	0	0	0	0	0
Finland	0.151	0.163	0.162	0.180	0.173	0.159
Former Czechoslovakia	—	—	—	—	—	—
Former Serbia and Montenegro	0.020	0.022	0.085	0.081	0.087	0.089
Former Yugoslavia	—	—	—	—	—	—
France	1.577	1.653	1.695	1.663	1.859	1.925
Germany	3.042	3.181	3.160	3.523	3.533	3.534
Germany, East	—	—	—	—	—	—
Germany, West	—	—	—	—	—	—
Gibraltar	0	0	0	0	0	0
Greece	0.075	0.074	0.079	0.090	0.098	0.104
Hungary	0.425	0.472	0.473	0.520	0.512	0.533
Iceland	0	0	0	0	0	0
Ireland	0.151	0.158	0.161	0.162	0.160	0.153
Italy	2.556	2.563	2.545	2.795	2.912	3.113
Luxembourg	0.030	0.031	0.046	0.047	0.053	0.052
Macedonia	0	0	0	0	0.004	0.004
Malta	0	0	0	0	0	0
Montenegro	—	—	—	—	—	—
Netherlands	1.543	1.581	1.580	1.587	1.620	1.562
Norway	0.152	0.176	0.203	0.238	0.303	0.211
Poland	0.440	0.451	0.445	0.491	0.519	0.539
Portugal	0.089	0.099	0.120	0.116	0.146	0.165
Romania	0.598	0.693	0.644	0.633	0.633	0.640

continues

TABLE 6.7 Continued
World Natural Gas Consumption, 2000–2005 [Quadrillion (10^{15}) Btu]

Serbia	—	—	—	—	—	—
Slovakia	0.255	0.272	0.259	0.250	0.243	0.234
Slovenia	0.037	0.038	0.035	0.037	0.039	0.041
Spain	0.671	0.723	0.827	0.937	1.087	1.284
Sweden	0.031	0.035	0.035	0.035	0.034	0.037
Switzerland	0.107	0.112	0.110	0.116	0.120	0.123
Turkey	0.548	0.590	0.650	0.782	0.825	1.004
United Kingdom	3.603	3.565	3.609	3.580	3.714	3.582
European totals	**17.820**	**18.416**	**18.654**	**19.696**	**20.511**	**20.938**
Eurasia						
Armenia	0.052	0.052	0.040	0.049	0.049	0.063
Azerbaijan	0.209	0.248	0.335	0.340	0.368	0.384
Belarus	0.718	0.659	0.620	0.658	0.601	0.743
Estonia	0.040	0.045	0.047	0.050	0.051	0.054
Former USSR	—	—	—	—	—	—
Georgia	0.045	0.043	0.054	0.038	0.045	0.055
Kazakhstan	0.514	0.529	0.551	0.583	0.582	1.126
Kyrgyzstan	0.071	0.075	0.044	0.027	0.034	0.027
Latvia	0.057	0.060	0.060	0.064	0.068	0.069
Lithuania	0.093	0.098	0.103	0.108	0.104	0.108
Moldova	0.080	0.077	0.083	0.089	0.082	0.092
Russia	14.497	14.787	14.946	15.689	16.439	16.573
Tajikistan	0.046	0.048	0.044	0.050	0.051	0.053
Turkmenistan	0.274	0.355	0.425	0.581	0.613	0.658
Ukraine	2.910	2.740	2.910	3.165	3.195	3.224
Uzbekistan	1.537	1.623	1.670	1.698	1.803	1.731
Eurasian totals	**21.141**	**21.439**	**21.934**	**23.189**	**24.084**	**24.960**
Middle East						
Bahrain	0.317	0.329	0.350	0.356	0.361	0.396
Iran	2.345	2.617	2.955	3.073	3.190	3.818
Iraq	0.116	0.102	0.087	0.041	0.065	0.091

continues

TABLE 6.7 Continued
World Natural Gas Consumption, 2000–2005 [Quadrillion (10¹⁵) Btu]

Israel	0.0004	0.0004	0.0004	0.001	0.028	0.026
Jordan	0.011	0.011	0.011	0.024	0.052	0.058
Kuwait	0.355	0.314	0.296	0.336	0.359	0.455
Lebanon	0	0	0	0	0	0
Oman	0.232	0.234	0.242	0.236	0.247	0.339
Qatar	0.557	0.405	0.411	0.451	0.556	0.691
Saudi Arabia	1.842	1.985	2.096	2.221	2.429	2.634
Syria	0.207	0.190	0.231	0.233	0.241	0.207
United Arab Emirates	1.162	1.201	1.347	1.401	1.487	1.525
Yemen	0	0	0	0	0	0
Middle Eastern totals	**7.144**	**7.389**	**8.027**	**8.373**	**9.013**	**10.240**
Africa						
Algeria	0.818	0.814	0.812	0.817	0.767	0.905
Angola	0.021	0.020	0.023	0.024	0.028	0.030
Benin	0	0	0	0	0	0
Botswana	0	0	0	0	0	0
Burkina Faso	0	0	0	0	0	0
Burundi	0	0	0	0	0	0
Cameroon	0	0	0	0	0	0
Cape Verde	0	0	0	0	0	0
Central African Republic	0	0	0	0	0	0
Chad	0	0	0	0	0	0
Comoros	0	0	0	0	0	0
Congo (Brazzaville)	0	0	0	0	0	0.004
Congo (Kinshasa)	0	0	0	0	0	0
Cote d'Ivoire (Ivory Coast)	0.050	0.050	0.050	0.050	0.048	0.048
Djibouti	0	0	0	0	0	0
Egypt	0.677	0.908	0.924	1.095	1.163	1.265
Equatorial Guinea	0.001	0.001	0.047	0.004	0.004	0.048
Eritrea	0	0	0	0	0	0
Ethiopia	0	0	0	0	0	0

continues

TABLE 6.7 Continued
World Natural Gas Consumption, 2000–2005 [Quadrillion (10^{15}) Btu]

Gabon	0.003	0.003	0.003	0.003	0.004	0.004
Gambia, The	0	0	0	0	0	0
Ghana	0	0	0	0	0	0
Guinea	0	0	0	0	0	0
Guinea-Bissau	0	0	0	0	0	0
Kenya	0	0	0	0	0	0
Lesotho	0	0	0	0	0	0
Liberia	0	0	0	0	0	0
Libya	0.192	0.199	0.206	0.176	0.219	0.216
Madagascar	0	0	0	0	0	0
Malawi	0	0	0	0	0	0
Mali	0	0	0	0	0	0
Mauritania	0	0	0	0	0	0
Mauritius	0	0	0	0	0	0
Morocco	0.002	0.002	0.002	0.002	0.002	0.002
Mozambique	0.002	0.002	0.002	0.003	0.003	0.007
Namibia	0	0	0	0	0	0
Niger	0	0	0	0	0	0
Nigeria	0.249	0.230	0.235	0.315	0.345	0.383
Reunion	0	0	0	0	0	0
Rwanda	0	0	0	0	0	0
Saint Helena	0	0	0	0	0	0
Sao Tome and Principe	0	0	0	0	0	0
Senegal	0.002	0.002	0.002	0.002	0.002	0.002
Seychelles	0	0	0	0	0	0
Sierra Leone	0	0	0	0	0	0
Somalia	0	0	0	0	0	0
South Africa	0.061	0.078	0.085	0.084	0.082	0.081
Sudan	0	0	0	0	0	0
Swaziland	0	0	0	0	0	0
Tanzania	0	0	0	0	0	0

continues

TABLE 6.7 Continued
World Natural Gas Consumption, 2000–2005 [Quadrillion (10¹⁵) Btu]

Togo	0	0	0	0	0	0
Tunisia	0.124	0.155	0.155	0.149	0.149	0.174
Uganda	0	0	0	0	0	0
Western Sahara	0	0	0	0	0	0
Zambia	0	0	0	0	0	0
Zimbabwe	0	0	0	0	0	0
African totals	**2.202**	**2.462**	**2.547**	**2.722**	**2.816**	**3.168**
Asia and Oceania						
Afghanistan	0.009	0.002	0.002	0.001	0.001	0.001
American Samoa	0	0	0	0	0	0
Australia	0.850	0.896	0.946	0.977	0.987	1.004
Bangladesh	0.335	0.356	0.384	0.420	0.453	0.484
Bhutan	0	0	0	0	0	0
Brunei	0.041	0.050	0.062	0.064	0.074	0.087
Burma	0.059	0.062	0.069	0.079	0.084	0.131
Cambodia	0	0	0	0	0	0
China	1.049	1.132	1.233	1.329	1.569	1.923
Cook Islands	0	0	0	0	0	0
East Timor	—	—	—	0	0	0
Fiji	0	0	0	0	0	0
French Polynesia	0	0	0	0	0	0
Guam	0	0	0	0	0	0
Hawaiian Trade Zone	—	—	—	—	—	—
Hong Kong	0.100	0.101	0.096	0.071	0.093	0.114
India	0.822	0.880	0.957	0.997	1.125	1.312
Indonesia	1.179	1.289	1.328	1.332	1.427	1.444
Japan	2.963	2.961	3.065	3.188	3.278	3.226
Kiribati	0	0	0	0	0	0
Korea, North	0	0	0	0	0	0
Korea, South	0.748	0.806	0.914	0.952	1.132	1.193
Laos	0	0	0	0	0	0

continues

TABLE 6.7 Continued
World Natural Gas Consumption, 2000–2005 [Quadrillion (10¹⁵) Btu]

Macau	0	0	0	0	0	0
Malaysia	0.760	0.958	1.043	1.176	1.269	1.235
Maldives	0	0	0	0	0	0
Mongolia	0	0	0	0	0	0
Nauru	0	0	0	0	0	0
Nepal	0	0	0	0	0	0
New Caledonia	0	0	0	0	0	0
New Zealand	0.221	0.236	0.220	0.170	0.153	0.144
Niue	0	0	0	0	0	0
Pakistan	0.799	0.723	0.755	0.831	0.904	1.016
Papua New Guinea	0.004	0.004	0.004	0.005	0.005	0.004
Philippines	0.0003	0.007	0.069	0.100	0.100	0.100
Samoa	0	0	0	0	0	0
Singapore	0.055	0.044	0.043	0.197	0.244	0.244
Solomon Islands	0	0	0	0	0	0
Sri Lanka	0	0	0	0	0	0
Taiwan	0.268	0.267	0.317	0.324	0.389	0.406
Thailand	0.688	0.818	0.903	0.995	1.031	1.123
Tonga	0	0	0	0	0	0
U.S. Pacific Islands	0	0	0	0	0	0
Vanuatu	0	0	0	0	0	0
Vietnam	0.042	0.048	0.083	0.099	0.110	0.146
Wake Island	0	0	0	0	0	0
Asian and Oceanian totals	**10.991**	**11.639**	**12.493**	**13.307**	**14.429**	**15.336**
World totals	**91.306**	**92.723**	**95.986**	**99.480**	**103.474**	**107.613**

Source: Energy Information Administration.

TABLE 6.8
World Coal Consumption, 2000–2005 [Quadrillion (10^{15}) Btu]

Region/Country	2000	2001	2002	2003	2004	2005
North America						
Bermuda	0	0	0	0	0	0
Canada	1.500	1.467	1.616	1.617	1.643	1.672
Greenland	0	0	0	0	0	0
Mexico	0.294	0.329	0.349	0.405	0.304	0.382
Saint Pierre and Miquelon	0	0	0	0	0	0
United States	22.645	21.944	21.965	22.371	22.604	22.829
North American totals	**24.439**	**23.740**	**23.930**	**24.393**	**24.551**	**24.883**
Central and South America						
Antarctica	0	0	0	0	0	0
Antigua and Barbuda	0	0	0	0	0	0
Argentina	0.024	0.022	0.019	0.021	0.024	0.037
Aruba	0	0	0	0	0	0
Bahamas, The	0	0	0	0	0	0
Barbados	0	0	0	0	0	0
Belize	0	0	0	0	0	0
Bolivia	0	0	0	0	0	0
Brazil	0.441	0.436	0.426	0.435	0.463	0.442
Cayman Islands	0	0	0	0	0	0
Chile	0.136	0.107	0.114	0.112	0.162	0.171
Colombia	0.101	0.090	0.065	0.102	0.081	0.113
Costa Rica	0.00001	0.00002	0.00002	0.00002	0.00002	0.0001
Cuba	0.001	0.001	0.001	0.001	0.001	0.001
Dominica	0	0	0	0	0	0
Dominican Republic	0.002	0.005	0.006	0.027	0.020	0.012
Ecuador	0	0	0	0	0	0
El Salvador	0.00002	0.00003	0.00003	0.00003	0.00003	0.0001
Falkland Islands	0	0	0	0	0	0
French Guiana	0	0	0	0	0	0

continues

TABLE 6.8 Continued
World Coal Consumption, 2000–2005 [Quadrillion (10¹⁵) Btu]

Grenada	0	0	0	0	0	0
Guadeloupe	0	0	0	0	0	0
Guatemala	0.006	0.005	0.010	0.009	0.012	0.010
Guyana	0	0	0	0	0	0
Haiti	0	0	0	0	0	0
Honduras	0.003	0.003	0.004	0.004	0.004	0.005
Jamaica	0.001	0.001	0.002	0.002	0.002	0.002
Martinique	0	0	0	0	0	0
Montserrat	0	0	0	0	0	0
Netherlands Antilles	0	0	0	0	0	0
Nicaragua	0	0	0.001	0.001	0.001	0
Panama	0.002	0.002	0.001	0	0.00005	0
Paraguay	0	0	0	0	0	0
Peru	0.026	0.023	0.032	0.031	0.032	0.038
Puerto Rico	0.004	0.004	0.011	0.035	0.033	0.034
Saint Kitts and Nevis	0	0	0	0	0	0
Saint Lucia	0	0	0	0	0	0
Saint Vincent/Grenadines	0	0	0	0	0	0
Suriname	0	0	0	0	0	0
Trinidad and Tobago	0	0	0	0	0	0
Turks and Caicos Islands	0	0	0	0	0	0
Uruguay	0.00002	0.00004	0.0001	0.00005	0.0001	0.0001
Venezuela	0.005	0.002	0.001	0.002	0	0.002
Virgin Islands, U.S.	0.007	0.007	0.007	0.007	0.007	0.008
Virgin Islands, British	0	0	0	0	0	0
Central and South American totals	**0.759**	**0.707**	**0.699**	**0.791**	**0.842**	**0.875**
Europe						
Albania	0.001	0.001	0.001	0.001	0.001	0.001
Austria	0.147	0.152	0.153	0.165	0.161	0.163
Belgium	0.351	0.323	0.254	0.238	0.232	0.208
Bosnia and Herzegovina	0.112	0.116	0.123	0.127	0.126	0.137

continues

TABLE 6.8 Continued
World Coal Consumption, 2000–2005 [Quadrillion (10¹⁵) Btu]

Bulgaria	0.263	0.294	0.267	0.299	0.291	0.283
Croatia	0.021	0.020	0.024	0.028	0.029	0.028
Cyprus	0.001	0.001	0.001	0.001	0.001	0.001
Czech Republic	0.792	0.804	0.765	0.775	0.711	0.748
Denmark	0.167	0.176	0.174	0.240	0.185	0.146
Faroe Islands	0	0	0	0	0	0
Finland	0.147	0.169	0.181	0.240	0.220	0.132
Former Czechoslovakia	—	—	—	—	—	—
Former Serbia and Montenegro	0.342	0.331	0.348	0.371	0.388	0.379
Former Yugoslavia	—	—	—	—	—	0
France	0.578	0.483	0.523	0.558	0.528	0.553
Germany	3.448	3.520	3.422	3.440	3.539	3.382
Germany, East	—	—	—	—	—	—
Germany, West	—	—	—	—	—	—
Gibraltar	0	0	0	0	0	0
Greece	0.377	0.388	0.375	0.379	0.375	0.365
Hungary	0.156	0.151	0.150	0.156	0.146	0.137
Iceland	0.004	0.004	0.004	0.004	0.004	0.004
Ireland	0.080	0.079	0.074	0.073	0.079	0.081
Italy	0.495	0.529	0.547	0.597	0.669	0.662
Luxembourg	0.005	0.004	0.004	0.003	0.004	0.003
Macedonia	0.056	0.060	0.054	0.055	0.055	0.054
Malta	0	0	0	0	0	0
Montenegro	—	—	—	—	—	—
Netherlands	0.318	0.331	0.334	0.342	0.346	0.329
Norway	0.045	0.039	0.034	0.033	0.039	0.126
Poland	2.342	2.177	2.172	2.256	2.263	2.192
Portugal	0.157	0.133	0.145	0.137	0.140	0.139
Romania	0.306	0.332	0.340	0.359	0.359	0.347
Serbia	—	—	—	—	—	—
Slovakia	0.167	0.174	0.167	0.181	0.177	0.168

continues

TABLE 6.8 Continued
World Coal Consumption, 2000–2005 [Quadrillion (10^{15}) Btu]

Slovenia	0.070	0.072	0.080	0.077	0.077	0.075
Spain	0.970	0.899	0.997	0.920	0.975	0.965
Sweden	0.088	0.098	0.099	0.093	0.102	0.092
Switzerland	0.006	0.006	0.006	0.006	0.006	0.006
Turkey	0.951	0.795	0.808	0.863	0.889	1.012
United Kingdom	1.404	1.584	1.457	1.577	1.518	1.550
European totals	**14.367**	**14.246**	**14.085**	**14.594**	**14.634**	**14.468**
Eurasia						
Armenia	0.002	0.003	0.002	0.002	0.002	0.001
Azerbaijan	0	0	0	0	0	0
Belarus	0.015	0.013	0.011	0.010	0.008	0.006
Estonia	0.112	0.112	0.109	0.129	0.130	0.124
Former USSR	—	—	—	—	—	—
Georgia	0.0005	0.001	0.001	0.001	0.0003	0.001
Kazakhstan	0.861	0.961	0.984	1.085	1.163	1.149
Kyrgyzstan	0.020	0.013	0.022	0.026	0.025	0.023
Latvia	0.002	0.003	0.003	0.002	0.002	0.003
Lithuania	0.004	0.003	0.006	0.007	0.007	0.008
Moldova	0.004	0.003	0.004	0.005	0.004	0.004
Russia	4.825	4.668	4.660	4.718	4.604	4.804
Tajikistan	0.001	0.001	0.001	0.001	0.001	0.002
Turkmenistan	0	0	0	0	0	0
Ukraine	1.427	1.381	1.394	1.541	1.393	1.373
Uzbekistan	0.037	0.039	0.039	0.027	0.038	0.043
Eurasian totals	**7.310**	**7.201**	**7.234**	**7.555**	**7.377**	**7.540**
Middle East						
Bahrain	0	0	0	0	0	0
Iran	0.048	0.044	0.048	0.046	0.040	0.046
Iraq	0	0	0	0	0	0
Israel	0.276	0.300	0.325	0.330	0.332	0.341
Jordan	0	0	0	0	0	0

continues

TABLE 6.8 Continued
World Coal Consumption, 2000–2005 [Quadrillion (10¹⁵) Btu]

Kuwait	0	0	0	0	0	0
Lebanon	0.006	0.006	0.006	0.006	0.006	0.006
Oman	0	0	0	0	0	0
Qatar	0	0	0	0	0	0
Saudi Arabia	0	0	0	0	0	0
Syria	0.0001	0.0001	0.0001	0.0001	0.0001	0.0001
United Arab Emirates	0	0	0	0	0	0
Yemen	0	0	0	0	0	0
Middle Eastern totals	**0.330**	**0.350**	**0.379**	**0.382**	**0.378**	**0.393**
Africa						
Algeria	0.020	0.021	0.026	0.028	0.027	0.026
Angola	0	0	0	0	0	0
Benin	0	0	0	0	0	0
Botswana	0.027	0.024	0.025	0.022	0.023	0.025
Burkina Faso	0	0	0	0	0	0
Burundi	0	0	0	0	0	0
Cameroon	0.00003	0	0	0	0	0
Cape Verde	0	0	0	0	0	0
Central African Republic	0	0	0	0	0	0
Chad	0	0	0	0	0	0
Comoros	0	0	0	0	0	0
Congo (Brazzaville)	0	0	0	0	0	0
Congo (Kinshasa)	0.006	0.006	0.006	0.007	0.007	0.007
Cote d'Ivoire (Ivory Coast)	0	0	0	0	0	0
Djibouti	0	0	0	0	0	0
Egypt	0.026	0.032	0.035	0.036	0.037	0.037
Equatorial Guinea	0	0	0	0	0	0
Eritrea	0	0	0	0	0	0
Ethiopia	0	0	0	0	0	0
Gabon	0	0	0	0	0	0
Gambia, The	0	0	0	0	0	0

continues

TABLE 6.8 Continued
World Coal Consumption, 2000–2005 [Quadrillion (10^{15}) Btu]

Ghana	0	0	0	0	0	0
Guinea	0	0	0	0	0	0
Guinea-Bissau	0	0	0	0	0	0
Kenya	0.002	0.002	0.003	0.002	0.003	0.003
Lesotho	0	0	0	0	0	0
Liberia	0	0	0	0	0	0
Libya	0	0	0	0	0	0
Madagascar	0.0003	0.0003	0.0003	0.0003	0.0003	0.0003
Malawi	0.0002	0.001	0.001	0.0001	0.0001	0.0001
Mali	0	0	0	0	0	0
Mauritania	0.0002	0.0002	0.0002	0.0002	0.0002	0.0002
Mauritius	0.007	0.009	0.008	0.007	0.009	0.008
Morocco	0.110	0.140	0.143	0.134	0.152	0.186
Mozambique	0	0	0.0002	0.001	0	0
Namibia	0.0001	0.0001	0.0001	0.0001	0.0001	0.0001
Niger	0.004	0.005	0.005	0.005	0.005	0.005
Nigeria	0.0002	0.0001	0.001	0.001	0.0001	0.0002
Reunion	0	0	0	0	0	0
Rwanda	0	0	0	0	0	0
Saint Helena	0	0	0	0	0	0
Sao Tome and Principe	0	0	0	0	0	0
Senegal	0	0	0	0	0.003	0.004
Seychelles	0	0	0	0	0	0
Sierra Leone	0	0	0	0	0	0
Somalia	0	0	0	0	0	0
South Africa	3.442	3.514	3.331	3.658	3.959	3.813
Sudan	0	0	0	0	0	0
Swaziland	0.008	0.007	0.007	0.007	0.007	0.007
Tanzania	0.002	0.002	0.002	0.001	0.002	0.002
Togo	0	0	0	0	0	0
Tunisia	0.003	0.003	0.002	0.001	0	0

continues

TABLE 6.8 Continued
World Coal Consumption, 2000–2005 [Quadrillion (10¹⁵) Btu]

Uganda	0	0	0	0	0	0
Western Sahara	0	0	0	0	0	0
Zambia	0.003	0.003	0.003	0.004	0.004	0.004
Zimbabwe	0.107	0.104	0.100	0.089	0.086	0.103
African totals	**3.768**	**3.874**	**3.698**	**4.003**	**4.323**	**4.230**
Asia and Oceania						
Afghanistan	0.002	0.002	0.002	0.002	0.002	0.002
American Samoa	0	0	0	0	0	0
Australia	2.086	2.208	2.247	2.204	2.273	2.516
Bangladesh	0.014	0.015	0.015	0.015	0.015	0.015
Bhutan	0.002	0.002	0.002	0.002	0.002	0.002
Brunei	0	0	0	0	0	0
Burma	0.001	0.002	0.002	0.003	0.004	0.004
Cambodia	0	0	0	0	0	0
China	23.974	25.286	28.418	34.662	41.409	46.905
Cook Islands	0	0	0	0	0	0
East Timor	—	—	—	NA	NA	NA
Fiji	0.0005	0.0004	0.0004	0.0004	0.0003	0.0003
French Polynesia	0	0	0	0	0	0
Guam	0	0	0	0	0	0
Hawaiian Trade Zone	—	—	—	—	—	—
Hong Kong	0.155	0.206	0.223	0.273	0.274	0.277
India	7.289	7.541	7.269	7.451	8.360	8.646
Indonesia	0.523	0.699	0.749	0.785	0.927	1.056
Japan	3.939	4.037	4.195	4.337	4.765	4.596
Kiribati	0	0	0	0	0	0
Korea, North	0.702	0.725	0.687	0.699	0.715	0.755
Korea, South	1.624	1.763	1.901	1.990	2.089	2.143
Laos	0.006	0.007	0.007	0.007	0.007	0.007
Macau	0.00003	0.00003	0.00003	0.00004	0.00004	0.00003
Malaysia	0.096	0.121	0.153	0.222	0.312	0.244

continues

TABLE 6.8 Continued
World Coal Consumption, 2000–2005 [Quadrillion (10^{15}) Btu]

Maldives	0	0	0	0	0	0
Mongolia	0.057	0.060	0.063	0.059	0.063	0.066
Nauru	0	0	0	0	0	0
Nepal	0.011	0.012	0.006	0.007	0.008	0.008
New Caledonia	0.007	0.007	0.007	0.008	0.007	0.008
New Zealand	0.049	0.068	0.068	0.100	0.079	0.086
Niue	0	0	0	0	0	0
Pakistan	0.084	0.092	0.105	0.137	0.176	0.169
Papua New Guinea	0	0	0	0	0	0
Philippines	0.210	0.219	0.205	0.199	0.216	0.254
Samoa	0	0	0	0	0	0
Singapore	0.000001	0.00001	0.00028	0.00036	0.00028	0.00008
Solomon Islands	0	0	0	0	0	0
Sri Lanka	0.0001	0.00003	0.0001	0.003	0.003	0.003
Taiwan	1.226	1.310	1.395	1.502	1.559	1.616
Thailand	0.325	0.371	0.383	0.403	0.446	0.486
Tonga	0	0	0	0	0	0
U.S. Pacific Islands	0	0	0	0	0	0
Vanuatu	0	0	0	0	0	0
Vietnam	0.182	0.209	0.230	0.243	0.348	0.339
Wake Island	0	0	0	0	0	0
Asian and Oceanian totals	**42.564**	**44.960**	**48.331**	**55.315**	**64.059**	**70.203**
World totals	**93.537**	**95.077**	**98.356**	**107.033**	**116.164**	**122.592**

Source: Energy Information Administration.

TABLE 6.9
World Net Hydroelectric Power Generation, 2000–2005
(Billion Kilowatt-hours)

Region/Country	2000	2001	2002	2003	2004	2005
North America						
Bermuda	0	0	0	0	0	0
Canada	354.92	329.99	347.05	334.07	337.43	359.88
Greenland	0	0	0	0	0	0
Mexico	32.80	28.22	24.70	19.68	24.95	27.46
Saint Pierre and Miquelon	0	0	0	0	0	0
United States	275.57	216.96	264.33	275.81	268.42	270.32
North American totals	**663.30**	**575.16**	**636.08**	**629.56**	**630.80**	**657.66**
Central and South America						
Antarctica	0	0	0	0	0	0
Antigua and Barbuda	0	0	0	0	0	0
Argentina	28.55	36.68	35.53	33.50	30.22	33.92
Aruba	0	0	0	0	0	0
Bahamas, The	0	0	0	0	0	0
Barbados	0	0	0	0	0	0
Belize	0.09	0.10	0.10	0.10	0.11	0.11
Bolivia	1.95	2.15	2.20	2.55	2.15	2.47
Brazil	301.36	265.20	283.23	302.56	317.59	334.08
Cayman Islands	0	0	0	0	0	0
Chile	19.55	21.55	23.18	23.10	23.19	23.80
Colombia	31.75	31.48	33.63	35.81	39.68	39.41
Costa Rica	5.62	5.60	5.87	5.86	6.42	6.50
Cuba	0.09	0.07	0.11	0.13	0.09	0.09
Dominica	0.03	0.03	0.03	0.03	0.03	0.04
Dominican Republic	0.75	0.55	0.87	1.19	1.57	1.88
Ecuador	7.53	7.00	7.45	7.11	7.34	6.81
El Salvador	1.16	1.15	1.13	1.45	1.37	1.65
Falkland Islands	0	0	0	0	0	0

continues

TABLE 6.9 Continued
World Net Hydroelectric Power Generation, 2000–2005
(Billion Kilowatt-hours)

French Guiana	0	0	0	0	0	0
Grenada	0	0	0	0	0	0
Guadeloupe	0	0	0	0	0	0
Guatemala	2.26	1.91	1.69	2.45	2.41	3.20
Guyana	0.01	0.01	0.01	0.01	0.01	0.01
Haiti	0.28	0.28	0.26	0.25	0.26	0.26
Honduras	2.24	2.35	2.43	2.15	2.33	1.79
Jamaica	0.11	0.06	0.09	0.12	0.13	0.15
Martinique	0	0	0	0	0	0
Montserrat	0	0	0	0	0	0
Netherlands Antilles	0	0	0	0	0	0
Nicaragua	0.21	0.20	0.30	0.29	0.32	0.43
Panama	3.38	2.47	3.37	2.80	3.74	3.69
Paraguay	52.96	44.86	47.73	51.25	51.41	50.65
Peru	16.01	17.43	17.86	18.35	17.35	19.76
Puerto Rico	0.15	0.13	0.11	0.10	0.14	0.14
Saint Kitts and Nevis	0	0	0	0	0	0
Saint Lucia	0	0	0	0	0	0
Saint Vincent/Grenadines	0.03	0.03	0.03	0.03	0.03	0.04
Suriname	1.32	1.33	1.35	1.36	1.37	1.40
Trinidad and Tobago	0	0	0	0	0	0
Turks and Caicos Islands	0	0	0	0	0	0
Uruguay	6.98	9.10	9.44	8.45	4.73	6.62
Venezuela	62.20	59.85	58.94	60.03	69.23	74.28
Virgin Islands, U.S.	0	0	0	0	0	0
Virgin Islands, British	0	0	0	0	0	0
Central and South American totals	**546.58**	**511.56**	**536.92**	**561.03**	**583.22**	**613.16**
Europe						
Albania	4.55	3.52	3.48	5.12	5.41	5.32
Austria	41.42	39.79	39.53	32.55	36.06	35.52

continues

TABLE 6.9 Continued
World Net Hydroelectric Power Generation, 2000–2005
(Billion Kilowatt-hours)

Belgium	0.46	0.44	0.36	0.25	0.31	0.29
Bosnia and Herzegovina	5.04	5.13	5.22	5.35	5.84	5.40
Bulgaria	2.92	2.15	2.68	3.27	3.33	4.68
Croatia	5.83	6.52	5.38	4.89	6.98	6.37
Cyprus	0	0	0	0	0	0
Czech Republic	1.74	2.03	2.47	1.37	2.00	2.36
Denmark	0.03	0.03	0.03	0.02	0.03	0.02
Faroe Islands	0.08	0.08	0.09	0.09	0.09	0.09
Finland	14.51	13.07	10.67	9.50	14.92	13.65
Former Czechoslovakia	—	—	—	—	—	—
Former Serbia and Montenegro	11.88	12.33	11.63	9.75	11.01	11.91
Former Yugoslavia	—	—	—	—	—	—
France	66.47	73.89	59.99	58.57	59.22	51.18
Germany	21.52	22.51	22.89	19.07	20.87	19.39
Germany, East	—	—	—	—	—	—
Germany, West	—	—	—	—	—	—
Gibraltar	0	0	0	0	0	0
Greece	3.66	2.08	2.77	4.72	4.63	4.97
Hungary	0.18	0.18	0.19	0.17	0.20	0.20
Iceland	6.29	6.51	6.91	7.02	7.06	6.95
Ireland	0.84	0.59	0.90	0.59	0.62	0.63
Italy	43.76	46.34	39.13	33.45	39.21	33.27
Luxembourg	0.12	0.13	0.11	0.08	0.11	0.10
Macedonia	1.16	0.62	0.75	1.36	1.47	1.48
Malta	0	0	0	0	0	0
Montenegro	—	—	—	—	—	—
Netherlands	0.14	0.12	0.11	0.07	0.09	0.09
Norway	137.53	117.12	128.12	104.56	107.77	134.44
Poland	2.09	2.30	2.26	1.65	2.06	2.18
Portugal	11.21	13.89	7.72	15.57	9.77	4.68

continues

TABLE 6.9 Continued
World Net Hydroelectric Power Generation, 2000–2005
(Billion Kilowatt-hours)

Romania	14.63	14.77	15.89	13.13	16.35	20.01
Serbia	—	—	—	—	—	—
Slovakia	4.57	4.88	5.22	3.45	4.06	4.59
Slovenia	3.77	3.74	3.27	2.92	4.03	3.41
Spain	29.27	40.62	22.69	40.64	31.24	19.36
Sweden	77.80	78.27	65.70	53.01	59.52	72.08
Switzerland	36.47	40.90	34.86	34.47	33.41	30.91
Turkey	30.57	23.77	33.35	34.98	45.62	39.17
United Kingdom	5.04	4.02	4.74	3.20	4.88	4.91
European totals	**585.52**	**582.32**	**539.08**	**504.78**	**538.18**	**539.57**
Eurasia						
Armenia	1.25	0.96	1.64	1.96	1.98	1.76
Azerbaijan	1.52	1.29	2.00	2.44	2.73	2.98
Belarus	0.03	0.03	0.03	0.03	0.03	0.04
Estonia	0.01	0.01	0.01	0.01	0.02	0.02
Former USSR	—	—	—	—	—	—
Georgia	5.80	5.49	6.70	6.46	5.99	6.17
Kazakhstan	7.46	8.00	8.80	8.54	7.98	7.78
Kyrgyzstan	13.55	12.31	10.68	13.38	13.95	14.12
Latvia	2.79	2.81	2.44	2.24	3.08	3.29
Lithuania	0.34	0.32	0.35	0.32	0.42	0.45
Moldova	0.32	0.32	0.32	0.33	0.33	0.31
Russia	163.72	174.09	162.55	156.14	176.01	172.86
Tajikistan	13.77	13.91	14.79	15.96	15.94	16.52
Turkmenistan	0.01	0.003	0.003	0.003	0.003	0.003
Ukraine	11.34	12.07	9.69	9.30	11.77	12.38
Uzbekistan	5.82	5.90	6.27	6.28	6.49	6.07
Eurasian totals	**227.70**	**237.49**	**226.27**	**223.41**	**246.71**	**244.73**
Middle East						
Bahrain	0	0	0	0	0	0
Iran	3.61	5.02	8.00	11.02	10.56	15.94

continues

TABLE 6.9 Continued
World Net Hydroelectric Power Generation, 2000–2005
(Billion Kilowatt-hours)

Iraq	0.61	0.61	0.51	0.43	0.49	0.51
Israel	0.03	0.01	0.02	0.03	0.03	0.03
Jordan	0.04	0.04	0.05	0.04	0.05	0.06
Kuwait	0	0	0	0	0	0
Lebanon	0.45	0.33	0.67	1.35	1.11	1.04
Oman	0	0	0	0	0	0
Qatar	0	0	0	0	0	0
Saudi Arabia	0	0	0	0	0	0
Syria	3.20	3.38	3.47	2.78	4.21	3.41
United Arab Emirates	0	0	0	0	0	0
Yemen	0	0	0	0	0	0
Middle Eastern totals	**7.93**	**9.40**	**12.72**	**15.65**	**16.44**	**20.98**
Africa						
Algeria	0.05	0.07	0.06	0.26	0.25	0.55
Angola	0.90	1.01	1.13	1.23	1.48	1.73
Benin	0.002	0.002	0.002	0.002	0.001	0.001
Botswana	0	0	0	0	0	0
Burkina Faso	0.10	0.05	0.06	0.10	0.10	0.10
Burundi	0.10	0.11	0.13	0.13	0.13	0.14
Cameroon	3.41	3.44	3.15	3.49	3.88	3.87
Cape Verde	0	0	0	0	0	0
Central African Republic	0.08	0.08	0.08	0.08	0.09	0.09
Chad	0	0	0	0	0	0
Comoros	0.002	0.002	0.002	0.002	0.002	0.002
Congo (Brazzaville)	0.29	0.34	0.39	0.34	0.39	0.35
Congo (Kinshasa)	5.94	5.82	6.02	6.36	6.76	7.32
Cote d'Ivoire (Ivory Coast)	1.75	1.78	1.71	1.81	1.73	1.42
Djibouti	0	0	0	0	0	0
Egypt	14.04	14.27	13.86	12.81	12.52	12.14
Equatorial Guinea	0.002	0.002	0.003	0.003	0.003	0.003

continues

TABLE 6.9 Continued
World Net Hydroelectric Power Generation, 2000–2005
(Billion Kilowatt-hours)

Eritrea	0	0	0	0	0	0
Ethiopia	1.63	1.97	2.00	2.26	2.50	2.82
Gabon	0.80	0.87	0.91	0.89	0.88	0.81
Gambia, The	0	0	0	0	0	0
Ghana	6.54	6.54	4.99	3.85	5.23	5.28
Guinea	0.41	0.43	0.43	0.44	0.44	0.42
Guinea-Bissau	0	0	0	0	0	0
Kenya	1.31	2.38	3.09	3.23	2.84	3.00
Lesotho	0.29	0.29	0.31	0.33	0.30	0.35
Liberia	0	0	0	0	0	0
Libya	0	0	0	0	0	0
Madagascar	0.54	0.53	0.54	0.61	0.64	0.65
Malawi	1.00	1.02	1.14	1.15	1.27	1.37
Mali	0.23	0.23	0.24	0.23	0.24	0.24
Mauritania	0.03	0.03	0.03	0.03	0.04	0.04
Mauritius	0.10	0.07	0.09	0.12	0.12	0.13
Morocco	0.71	0.86	0.84	1.44	1.60	1.41
Mozambique	8.75	8.72	12.55	10.76	11.55	13.13
Namibia	1.37	1.36	1.45	1.50	1.59	1.64
Niger	0	0	0	0	0	0
Nigeria	5.69	5.97	8.32	7.53	8.19	7.87
Reunion	0.55	0.56	0.57	0.57	0.57	0.58
Rwanda	0.11	0.09	0.10	0.10	0.09	0.09
Saint Helena	0	0	0	0	0	0
Sao Tome and Principe	0.01	0.01	0.01	0.01	0.01	0.01
Senegal	0	0	0.19	0.33	0.29	0.26
Seychelles	0	0	0	0	0	0
Sierra Leone	0	0	0	0	0	0
Somalia	0	0	0	0	0	0
South Africa	1.34	2.06	2.36	0.78	0.72	0.90
Sudan	1.17	1.22	1.27	1.15	1.05	1.23

continues

TABLE 6.9 Continued
World Net Hydroelectric Power Generation, 2000–2005
(Billion Kilowatt-hours)

Swaziland	0.20	0.19	0.19	0.19	0.20	0.20
Tanzania	2.12	2.58	2.69	2.52	2.33	1.76
Togo	0.20	0.09	0.17	0.24	0.08	0.07
Tunisia	0.06	0.05	0.06	0.16	0.15	0.14
Uganda	1.55	1.55	1.67	1.75	1.89	1.98
Western Sahara	0	0	0	0	0	0
Zambia	7.67	7.81	8.02	8.17	8.38	8.79
Zimbabwe	3.23	2.97	3.79	5.31	5.47	5.78
African totals	**74.26**	**77.43**	**84.61**	**82.27**	**85.96**	**88.66**
Asia and Oceania						
Afghanistan	0.31	0.29	0.56	0.63	0.56	0.53
American Samoa	0	0	0	0	0	0
Australia	16.61	16.25	15.70	15.89	15.52	15.53
Bangladesh	0.94	0.98	1.06	1.12	1.21	1.28
Bhutan	1.79	1.88	1.88	1.88	1.93	2.00
Brunei	0	0	0	0	0	0
Burma	1.87	1.80	2.21	2.23	2.38	2.97
Cambodia	0.05	0.03	0.03	0.04	0.03	0.04
China	240.70	258.50	271.82	278.52	327.68	396.99
Cook Islands	0	0	0	0	0	0
East Timor	—	—	—	NA	NA	NA
Fiji	0.55	0.58	0.63	0.67	0.67	0.70
French Polynesia	0.12	0.11	0.09	0.09	0.09	0.09
Guam	0	0	0	0	0	0
Hawaiian Trade Zone	—	—	—	—	—	—
Hong Kong	0	0	0	0	0	0
India	73.72	72.96	63.46	74.59	83.88	99.00
Indonesia	9.92	11.54	9.84	8.99	9.58	10.65
Japan	86.38	83.32	81.55	93.66	93.12	77.43
Kiribati	0	0	0	0	0	0

continues

TABLE 6.9 Continued
World Net Hydroelectric Power Generation, 2000–2005
(Billion Kilowatt-hours)

Korea, North	10.10	10.49	10.51	11.60	12.38	13.00
Korea, South	3.97	4.11	3.20	4.85	4.29	3.64
Laos	1.53	1.51	1.52	1.28	1.37	1.66
Macau	0	0	0	0	0	0
Malaysia	7.34	6.37	5.25	5.69	5.77	5.73
Maldives	0	0	0	0	0	0
Mongolia	0	0	0	0	0	0
Nauru	0	0	0	0	0	0
Nepal	1.62	1.83	2.10	2.24	2.32	2.39
New Caledonia	0.44	0.36	0.32	0.32	0.32	0.33
New Zealand	24.14	21.24	24.98	23.46	26.93	23.24
Niue	0	0	0	0	0	0
Pakistan	17.02	18.75	22.13	26.68	25.41	30.55
Papua New Guinea	0.92	0.92	0.92	0.92	0.93	0.93
Philippines	7.72	7.03	6.96	7.79	8.51	8.30
Samoa	0.04	0.05	0.04	0.04	0.04	0.04
Singapore	0	0	0	0	0	0
Solomon Islands	0	0	0	0	0	0
Sri Lanka	3.17	3.08	2.67	3.28	2.93	3.42
Taiwan	8.71	9.11	6.30	6.83	6.50	7.83
Thailand	5.97	6.24	7.40	7.23	5.98	5.74
Tonga	0	0	0	0	0	0
U.S. Pacific Islands	0.03	0.03	0.03	0.03	0.03	0.03
Vanuatu	0	0	0	0	0	0
Vietnam	14.41	18.03	18.02	18.80	17.49	21.24
Wake Island	0	0	0	0	0	0
Asian and Oceanian totals	**540.06**	**557.39**	**561.15**	**599.32**	**657.85**	**735.27**
World totals	**2,645.36**	**2,550.74**	**2,596.82**	**2,616.02**	**2,759.16**	**2,900.03**

Source: Energy Information Administration.

TABLE 6.10
World Net Nuclear Electric Power Generation, 2000–2005
(Billion Kilowatt-hours)

Region/Country	2000	2001	2002	2003	2004	2005
North America						
Bermuda	0	0	0	0	0	0
Canada	69.16	72.86	71.75	71.15	85.87	87.44
Greenland	0	0	0	0	0	0
Mexico	7.81	8.29	9.26	9.98	8.73	10.27
Saint Pierre and Miquelon	0	0	0	0	0	0
United States	753.89	768.83	780.06	763.73	788.53	781.99
North American totals	**830.86**	**849.97**	**861.07**	**844.85**	**883.13**	**879.69**
Central and South America						
Antarctica	0	0	0	0	0	0
Antigua and Barbuda	0	0	0	0	0	0
Argentina	5.99	6.54	5.39	7.03	7.31	6.37
Aruba	0	0	0	0	0	0
Bahamas, The	0	0	0	0	0	0
Barbados	0	0	0	0	0	0
Belize	0	0	0	0	0	0
Bolivia	0	0	0	0	0	0
Brazil	4.94	14.27	13.84	13.40	11.60	9.90
Cayman Islands	0	0	0	0	0	0
Chile	0	0	0	0	0	0
Colombia	0	0	0	0	0	0
Costa Rica	0	0	0	0	0	0
Cuba	0	0	0	0	0	0
Dominica	0	0	0	0	0	0
Dominican Republic	0	0	0	0	0	0
Ecuador	0	0	0	0	0	0
El Salvador	0	0	0	0	0	0
Falkland Islands	0	0	0	0	0	0

continues

TABLE 6.10 Continued
World Net Nuclear Electric Power Generation, 2000–2005
(Billion Kilowatt-hours)

French Guiana	0	0	0	0	0	0
Grenada	0	0	0	0	0	0
Guadeloupe	0	0	0	0	0	0
Guatemala	0	0	0	0	0	0
Guyana	0	0	0	0	0	0
Haiti	0	0	0	0	0	0
Honduras	0	0	0	0	0	0
Jamaica	0	0	0	0	0	0
Martinique	0	0	0	0	0	0
Montserrat	0	0	0	0	0	0
Netherlands Antilles	0	0	0	0	0	0
Nicaragua	0	0	0	0	0	0
Panama	0	0	0	0	0	0
Paraguay	0	0	0	0	0	0
Peru	0	0	0	0	0	0
Puerto Rico	0	0	0	0	0	0
Saint Kitts and Nevis	0	0	0	0	0	0
Saint Lucia	0	0	0	0	0	0
Saint Vincent/Grenadines	0	0	0	0	0	0
Suriname	0	0	0	0	0	0
Trinidad and Tobago	0	0	0	0	0	0
Turks and Caicos Islands	0	0	0	0	0	0
Uruguay	0	0	0	0	0	0
Venezuela	0	0	0	0	0	0
Virgin Islands, U.S.	0	0	0	0	0	0
Virgin Islands, British	0	0	0	0	0	0
Central and South American totals	**10.93**	**20.81**	**19.23**	**20.43**	**18.91**	**16.27**
Europe						
Albania	0	0	0	0	0	0
Austria	0	0	0	0	0	0

continues

TABLE 6.10 Continued
World Net Nuclear Electric Power Generation, 2000–2005
(Billion Kilowatt-hours)

Belgium	45.75	44.03	44.99	45.01	44.95	45.22
Bosnia and Herzegovina	0	0	0	0	0	0
Bulgaria	17.27	18.24	20.22	16.04	15.60	17.34
Croatia	0	0	0	0	0	0
Cyprus	0	0	0	0	0	0
Czech Republic	12.91	14.01	17.80	24.58	25.01	23.49
Denmark	0	0	0	0	0	0
Faroe Islands	0	0	0	0	0	0
Finland	21.36	21.63	21.18	21.60	21.58	22.10
Former Czechoslovakia	—	—	—	—	—	—
Former Serbia and Montenegro	0	0	0	0	0	0
Former Yugoslavia	—	—	—	—	—	—
France	394.40	400.02	414.92	419.02	425.83	428.95
Germany	161.13	162.74	156.60	156.81	158.71	154.85
Germany, East	—	—	—	—	—	—
Germany, West	—	—	—	—	—	—
Gibraltar	0	0	0	0	0	0
Greece	0	0	0	0	0	0
Hungary	13.47	13.42	13.26	10.46	11.32	13.14
Iceland	0	0	0	0	0	0
Ireland	0	0	0	0	0	0
Italy	0	0	0	0	0	0
Luxembourg	0	0	0	0	0	0
Macedonia	0	0	0	0	0	0
Malta	0	0	0	0	0	0
Montenegro	—	—	—	—	—	—
Netherlands	3.73	3.78	3.72	3.82	3.63	3.80
Norway	0	0	0	0	0	0
Poland	0	0	0	0	0	0
Portugal	0	0	0	0	0	0

continues

TABLE 6.10 Continued
World Net Nuclear Electric Power Generation, 2000–2005
(Billion Kilowatt-hours)

Romania	5.23	5.04	5.11	4.54	5.27	5.28
Serbia	—	—	—	—	—	—
Slovakia	15.67	16.25	17.06	16.97	16.18	16.84
Slovenia	4.55	5.04	5.31	4.96	5.21	5.61
Spain	59.10	60.52	59.87	58.78	60.43	54.66
Sweden	54.45	68.50	64.20	64.04	73.61	68.63
Switzerland	25.12	25.47	25.87	26.11	25.61	22.17
Turkey	0	0	0	0	0	0
United Kingdom	80.81	85.38	83.64	84.25	73.68	75.17
European totals	**914.94**	**944.07**	**953.74**	**957.00**	**966.62**	**957.27**
Eurasia						
Armenia	1.84	1.99	2.09	1.82	2.21	2.50
Azerbaijan	0	0	0	0	0	0
Belarus	0	0	0	0	0	0
Estonia	0	0	0	0	0	0
Former USSR	—	—	—	—	—	—
Georgia	0	0	0	0	0	0
Kazakhstan	0	0	0	0	0	0
Kyrgyzstan	0	0	0	0	0	0
Latvia	0	0	0	0	0	0
Lithuania	8.00	10.79	13.44	14.71	14.35	9.82
Moldova	0	0	0	0	0	0
Russia	122.46	125.36	134.14	141.17	137.47	140.22
Tajikistan	0	0	0	0	0	0
Turkmenistan	0	0	0	0	0	0
Ukraine	71.06	71.67	73.38	76.70	82.69	83.29
Uzbekistan	0	0	0	0	0	0
Eurasian totals	**203.35**	**209.81**	**223.05**	**234.40**	**236.71**	**235.83**
Middle East						
Bahrain	0	0	0	0	0	0
Iran	0	0	0	0	0	0

continues

TABLE 6.10 Continued
World Net Nuclear Electric Power Generation, 2000–2005
(Billion Kilowatt-hours)

Iraq	0	0	0	0	0	0
Israel	0	0	0	0	0	0
Jordan	0	0	0	0	0	0
Kuwait	0	0	0	0	0	0
Lebanon	0	0	0	0	0	0
Oman	0	0	0	0	0	0
Qatar	0	0	0	0	0	0
Saudi Arabia	0	0	0	0	0	0
Syria	0	0	0	0	0	0
United Arab Emirates	0	0	0	0	0	0
Yemen	0	0	0	0	0	0
Middle Eastern totals	**0**	**0**	**0**	**0**	**0**	**0**
Africa						
Algeria	0	0	0	0	0	0
Angola	0	0	0	0	0	0
Benin	0	0	0	0	0	0
Botswana	0	0	0	0	0	0
Burkina Faso	0	0	0	0	0	0
Burundi	0	0	0	0	0	0
Cameroon	0	0	0	0	0	0
Cape Verde	0	0	0	0	0	0
Central African Republic	0	0	0	0	0	0
Chad	0	0	0	0	0	0
Comoros	0	0	0	0	0	0
Congo (Brazzaville)	0	0	0	0	0	0
Congo (Kinshasa)	0	0	0	0	0	0
Cote d'Ivoire (Ivory Coast)	0	0	0	0	0	0
Djibouti	0	0	0	0	0	0
Egypt	0	0	0	0	0	0
Equatorial Guinea	0	0	0	0	0	0

continues

TABLE 6.10 Continued
World Net Nuclear Electric Power Generation, 2000–2005
(Billion Kilowatt-hours)

Eritrea	0	0	0	0	0	0
Ethiopia	0	0	0	0	0	0
Gabon	0	0	0	0	0	0
Gambia, The	0	0	0	0	0	0
Ghana	0	0	0	0	0	0
Guinea	0	0	0	0	0	0
Guinea-Bissau	0	0	0	0	0	0
Kenya	0	0	0	0	0	0
Lesotho	0	0	0	0	0	0
Liberia	0	0	0	0	0	0
Libya	0	0	0	0	0	0
Madagascar	0	0	0	0	0	0
Malawi	0	0	0	0	0	0
Mali	0	0	0	0	0	0
Mauritania	0	0	0	0	0	0
Mauritius	0	0	0	0	0	0
Morocco	0	0	0	0	0	0
Mozambique	0	0	0	0	0	0
Namibia	0	0	0	0	0	0
Niger	0	0	0	0	0	0
Nigeria	0	0	0	0	0	0
Reunion	0	0	0	0	0	0
Rwanda	0	0	0	0	0	0
Saint Helena	0	0	0	0	0	0
Sao Tome and Principe	0	0	0	0	0	0
Senegal	0	0	0	0	0	0
Seychelles	0	0	0	0	0	0
Sierra Leone	0	0	0	0	0	0
Somalia	0	0	0	0	0	0
South Africa	13.01	10.72	11.99	12.66	14.28	12.24
Sudan	0	0	0	0	0	0

continues

TABLE 6.10 Continued
World Net Nuclear Electric Power Generation, 2000–2005
(Billion Kilowatt-hours)

Swaziland	0	0	0	0	0	0
Tanzania	0	0	0	0	0	0
Togo	0	0	0	0	0	0
Tunisia	0	0	0	0	0	0
Uganda	0	0	0	0	0	0
Western Sahara	0	0	0	0	0	0
Zambia	0	0	0	0	0	0
Zimbabwe	0	0	0	0	0	0
African totals	**13.01**	**10.72**	**11.99**	**12.66**	**14.28**	**12.24**
Asia and Oceania						
Afghanistan	0	0	0	0	0	0
American Samoa	0	0	0	0	0	0
Australia	0	0	0	0	0	0
Bangladesh	0	0	0	0	0	0
Bhutan	0	0	0	0	0	0
Brunei	0	0	0	0	0	0
Burma	0	0	0	0	0	0
Cambodia	0	0	0	0	0	0
China	15.90	16.60	25.17	41.66	47.95	50.33
Cook Islands	0	0	0	0	0	0
East Timor	—	—	—	0	0	0
Fiji	0	0	0	0	0	0
French Polynesia	0	0	0	0	0	0
Guam	0	0	0	0	0	0
Hawaiian Trade Zone	—	—	—	—	—	—
Hong Kong	0	0	0	0	0	0
India	14.06	18.23	17.76	16.37	15.04	15.73
Indonesia	0	0	0	0	0	0
Japan	305.95	303.87	280.34	228.01	268.32	278.39
Kiribati	0	0	0	0	0	0

continues

TABLE 6.10 Continued
World Net Nuclear Electric Power Generation, 2000–2005
(Billion Kilowatt-hours)

Korea, North	0	0	0	0	0	0
Korea, South	103.52	106.53	113.15	123.19	124.18	139.44
Laos	0	0	0	0	0	0
Macau	0	0	0	0	0	0
Malaysia	0	0	0	0	0	0
Maldives	0	0	0	0	0	0
Mongolia	0	0	0	0	0	0
Nauru	0	0	0	0	0	0
Nepal	0	0	0	0	0	0
New Caledonia	0	0	0	0	0	0
New Zealand	0	0	0	0	0	0
Niue	0	0	0	0	0	0
Pakistan	0.38	1.98	1.80	1.81	1.93	2.41
Papua New Guinea	0	0	0	0	0	0
Philippines	0	0	0	0	0	0
Samoa	0	0	0	0	0	0
Singapore	0	0	0	0	0	0
Solomon Islands	0	0	0	0	0	0
Sri Lanka	0	0	0	0	0	0
Taiwan	37.00	34.09	38.01	37.37	37.94	37.97
Thailand	0	0	0	0	0	0
Tonga	0	0	0	0	0	0
U.S. Pacific Islands	0	0	0	0	0	0
Vanuatu	0	0	0	0	0	0
Vietnam	0	0	0	0	0	0
Wake Island	0	0	0	0	0	0
Asian and Oceanian totals	**476.80**	**481.30**	**476.22**	**448.41**	**495.36**	**524.27**
World totals	**2,449.89**	**2,516.67**	**2,545.30**	**2,517.76**	**2,615.01**	**2,625.57**

Source: Energy Information Administration.

TABLE 6.11

**World Net Geothermal, Solar, Wind, and Wood and Waste Electric Power Consumption,
2000–2005 (Billion Kilowatt-hours)**

Region/Country	2000	2001	2002	2003	2004	2005
North America						
Bermuda	0	0	0	0	0	0
Canada	7.98	8.42	8.91	9.32	9.20	10.05
Greenland	0	0	0	0	0	0
Mexico	6.07	5.81	5.62	8.34	8.68	9.45
Saint Pierre and Miquelon	0	0	0	0	0	0
United States	85.70	82.67	92.64	93.53	97.09	99.68
North American totals	**99.75**	**96.91**	**107.17**	**111.19**	**114.97**	**119.17**
Central and South America						
Antarctica	0	0	0	0	0	0
Antigua and Barbuda	0	0	0	0	0	0
Argentina	0.68	0.65	0.91	1.08	1.20	1.31
Aruba	0	0	0	0	0	0
Bahamas, The	0	0	0	0	0	0
Barbados	0	0	0	0	0	0
Belize	0	0	0	0	0	0
Bolivia	0.06	0.07	0.08	0.09	0.09	0.16
Brazil	11.80	13.26	14.56	15.97	17.21	18.33
Cayman Islands	0	0	0	0	0	0
Chile	0.67	1.52	1.47	1.68	1.48	0.86
Colombia	0.52	0.52	0.52	0.52	0.52	0.53
Costa Rica	1.13	1.14	1.37	1.45	1.47	1.60
Cuba	0.70	0.73	0.75	0.76	0.61	0.50
Dominica	0	0	0	0	0	0
Dominican Republic	0.04	0.04	0.04	0.04	0.07	0.06
Ecuador	0	0	0	0	0	0
El Salvador	0.77	0.94	0.96	0.99	0.97	1.00
Falkland Islands	0	0	0	0	0	0
French Guiana	0	0	0	0	0	0

continues

TABLE 6.11 Continued
**World Net Geothermal, Solar, Wind, and Wood and Waste Electric Power Consumption,
2000–2005 (Billion Kilowatt-hours)**

Grenada	0	0	0	0	0	0
Guadeloupe	0	0	0	0	0	0
Guatemala	0.81	0.81	0.81	0.81	0.83	0.86
Guyana	0	0	0	0	0	0
Haiti	0	0	0	0	0	0
Honduras	0.001	0.01	0.01	0.02	0.02	0.07
Jamaica	0.10	0.10	0.10	0.10	0.10	0.10
Martinique	0	0	0	0	0	0
Montserrat	0	0	0	0	0	0
Netherlands Antilles	0	0	0	0	0	0
Nicaragua	0.19	0.24	0.32	0.36	0.36	0.41
Panama	0.02	0.02	0.02	0.02	0.02	0.02
Paraguay	0	0	0	0	0	0
Peru	0.15	0.16	0.18	0.20	0.15	0.15
Puerto Rico	0	0	0	0	0	0
Saint Kitts and Nevis	0	0	0	0	0	0
Saint Lucia	0	0	0	0	0	0
Saint Vincent/Grenadines	0	0	0	0	0	0
Suriname	0	0	0	0	0	0
Trinidad and Tobago	0.02	0.03	0.03	0.01	0.02	0.02
Turks and Caicos Islands	0	0	0	0	0	0
Uruguay	0.03	0.03	0.03	0.03	0.03	0.04
Venezuela	0	0	0	0	0	0
Virgin Islands, U.S.	0	0	0	0	0	0
Virgin Islands, British	0	0	0	0	0	0
Central and South American totals	**17.68**	**20.26**	**22.16**	**24.09**	**25.14**	**26.00**
Europe						
Albania	0	0	0	0	0	0
Austria	1.75	1.98	1.98	2.28	3.10	3.73
Belgium	1.28	1.54	1.63	1.63	1.99	2.36

continues

TABLE 6.11 Continued
World Net Geothermal, Solar, Wind, and Wood and Waste Electric Power Consumption, 2000–2005 (Billion Kilowatt-hours)

Bosnia and Herzegovina	0	0	0	0	0	0
Bulgaria	0.01	0.01	0.01	0.01	0.02	0.02
Croatia	0.001	0.001	0	0	0	0
Cyprus	0	0	0	0	0	0
Czech Republic	0.69	0.68	0.66	0.48	0.69	0.72
Denmark	5.83	6.11	7.02	8.31	9.65	10.07
Faroe Islands	0	0	0	0	0	0
Finland	8.51	8.32	9.69	9.75	10.21	9.35
Former Czechoslovakia	—	—	—	—	—	—
Former Serbia and Montenegro	0	0	0	0	0	0
Former Yugoslavia	—	—	—	—	—	—
France	4.20	4.50	5.16	5.53	5.84	6.26
Germany	18.56	21.85	27.54	32.20	39.99	42.85
Germany, East	—	—	—	—	—	—
Germany, West	—	—	—	—	—	—
Gibraltar	0	0	0	0	0	0
Greece	0.58	0.89	0.84	1.20	1.32	1.42
Hungary	0.11	0.12	0.07	0.19	0.72	1.65
Iceland	1.26	1.38	1.36	1.34	1.41	1.58
Ireland	0.32	0.41	0.45	0.51	0.73	1.18
Italy	7.58	8.66	9.85	11.59	13.20	14.22
Luxembourg	0.08	0.08	0.08	0.09	0.12	0.14
Macedonia	0	0	0	0	0	0
Malta	0	0	0	0	0	0
Montenegro	—	—	—	—	—	—
Netherlands	4.14	4.34	5.07	5.22	6.35	8.64
Norway	0.32	0.40	0.40	0.65	0.70	0.89
Poland	0.53	0.74	0.90	0.82	1.26	1.94
Portugal	1.71	1.86	2.08	2.15	2.58	3.64

continues

TABLE 6.11 Continued
World Net Geothermal, Solar, Wind, and Wood and Waste Electric Power Consumption,
2000–2005 (Billion Kilowatt-hours)

Romania	0	0	0.003	0.003	0.004	0.01
Serbia	—	—	—	—	—	—
Slovakia	0.03	0.15	0.15	0.11	0.04	0.06
Slovenia	0.07	0.07	0.10	0.12	0.12	0.11
Spain	6.50	8.31	11.63	14.85	18.49	23.17
Sweden	4.56	4.28	4.79	5.24	8.41	8.83
Switzerland	1.53	1.44	1.47	1.83	1.91	2.02
Turkey	0.31	0.32	0.31	0.25	0.24	0.26
United Kingdom	5.01	5.72	6.54	9.73	11.61	15.02
European totals	**75.48**	**84.15**	**99.81**	**116.07**	**140.70**	**160.11**
Eurasia						
Armenia	0	0	0	0	0	0
Azerbaijan	0	0	0	0	0	0
Belarus	0	0	0	0	0	0
Estonia	0.01	0.01	0.02	0.03	0.04	0.09
Former USSR	—	—	—	—	—	—
Georgia	0	0	0	0	0	0
Kazakhstan	0	0	0	0	0	0
Kyrgyzstan	0	0	0	0	0	0
Latvia	0.004	0.003	0.01	0.07	0.08	0.09
Lithuania	0	0	0	0	0	0
Moldova	0	0	0	0	0	0
Russia	2.47	2.81	2.82	2.04	2.11	2.91
Tajikistan	0	0	0	0	0	0
Turkmenistan	0	0	0	0	0	0
Ukraine	0.01	0.02	0.02	0.03	0.03	0.04
Uzbekistan	0	0	0	0	0	0
Eurasian totals	**2.49**	**2.84**	**2.87**	**2.16**	**2.25**	**3.12**
Middle East						
Bahrain	0	0	0	0	0	0
Iran	0	0	0	0	0	0

continues

TABLE 6.11 Continued
**World Net Geothermal, Solar, Wind, and Wood and Waste Electric Power Consumption,
2000–2005 (Billion Kilowatt-hours)**

Iraq	0	0	0	0	0	0
Israel	0	0.01	0.01	0.01	0.01	0.01
Jordan	0.003	0.003	0.003	0.003	0.003	0.003
Kuwait	0	0	0	0	0	0
Lebanon	0	0	0	0	0	0
Oman	0	0	0	0	0	0
Qatar	0	0	0	0	0	0
Saudi Arabia	0	0	0	0	0	0
Syria	0	0	0	0	0	0
United Arab Emirates	0	0	0	0	0	0
Yemen	0	0	0	0	0	0
Middle Eastern totals	**0.003**	**0.01**	**0.01**	**0.02**	**0.01**	**0.01**
Africa						
Algeria	0	0	0	0	0	0
Angola	0	0	0	0	0	0
Benin	0	0	0	0	0	0
Botswana	0	0	0	0	0	0
Burkina Faso	0	0	0	0	0	0
Burundi	0	0	0	0	0	0
Cameroon	0	0	0	0	0	0
Cape Verde	0	0	0	0	0	0
Central African Republic	0	0	0	0	0	0
Chad	0	0	0	0	0	0
Comoros	0	0	0	0	0	0
Congo (Brazzaville)	0	0	0	0	0	0
Congo (Kinshasa)	0	0	0	0	0	0
Cote d'Ivoire (Ivory Coast)	0	0	0	0	0	0
Djibouti	0	0	0	0	0	0
Egypt	0.13	0.22	0.20	0.37	0.52	0.55
Equatorial Guinea	0	0	0	0	0	0
Eritrea	0.001	0.001	0.001	0.001	0.001	0.001

continues

TABLE 6.11 Continued
World Net Geothermal, Solar, Wind, and Wood and Waste Electric Power Consumption,
2000–2005 (Billion Kilowatt-hours)

Ethiopia	0.02	0.01	0.001	0.001	0.001	0.001
Gabon	0.01	0.01	0.01	0.01	0.01	0.01
Gambia, The	0	0	0	0	0	0
Ghana	0	0	0	0	0	0
Guinea	0	0	0	0	0	0
Guinea-Bissau	0	0	0	0	0	0
Kenya	0.41	0.46	0.26	0.65	0.87	0.84
Lesotho	0	0	0	0	0	0
Liberia	0	0	0	0	0	0
Libya	0	0	0	0	0	0
Madagascar	0	0	0	0	0	0
Malawi	0	0	0	0	0	0
Mali	0	0	0	0	0	0
Mauritania	0	0	0	0	0	0
Mauritius	0	0	0	0	0	0
Morocco	0.06	0.21	0.19	0.20	0.20	0.21
Mozambique	0	0	0	0	0	0
Namibia	0	0	0	0	0	0
Niger	0	0	0	0	0	0
Nigeria	0	0	0	0	0	0
Reunion	0	0	0	0	0	0
Rwanda	0	0	0	0	0	0
Saint Helena	0	0	0	0	0	0
Sao Tome and Principe	0	0	0	0	0	0
Senegal	0.002	0.002	0.002	0.002	0.003	0.004
Seychelles	0	0	0	0	0	0
Sierra Leone	0	0	0	0	0	0
Somalia	0	0	0	0	0	0
South Africa	0.29	0.29	0.30	0.30	0.30	0.30
Sudan	0	0	0	0	0	0
Swaziland	0	0	0	0	0	0

continues

TABLE 6.11 Continued
**World Net Geothermal, Solar, Wind, and Wood and Waste Electric Power Consumption,
2000–2005 (Billion Kilowatt-hours)**

Tanzania	0	0	0	0	0	0
Togo	0	0	0	0	0	0
Tunisia	0.02	0.02	0.03	0.03	0.04	0.04
Uganda	0	0	0	0	0	0
Western Sahara	0	0	0	0	0	0
Zambia	0	0	0	0	0	0
Zimbabwe	0	0	0	0	0	0
African totals	**0.95**	**1.21**	**1.00**	**1.56**	**1.96**	**1.96**
Asia and Oceania						
Afghanistan	0	0	0	0	0	0
American Samoa	0	0	0	0	0	0
Australia	1.07	1.50	1.83	2.06	2.41	2.78
Bangladesh	0	0	0	0	0	0
Bhutan	0	0	0	0	0	0
Brunei	0	0	0	0	0	0
Burma	0	0	0	0	0	0
Cambodia	0	0	0	0	0	0
China	2.30	2.32	2.33	2.35	2.37	2.38
Cook Islands	0	0	0	0	0	0
East Timor	—	—	—	NA	NA	NA
Fiji	0	0	0	0	0	0
French Polynesia	0	0	0	0	0	0
Guam	0	0	0	0	0	0
Hawaiian Trade Zone	—	—	—	—	—	—
Hong Kong	0	0	0	0	0	0
India	2.88	3.86	4.31	5.20	6.08	7.68
Indonesia	4.63	5.73	5.93	6.00	6.32	6.27
Japan	18.47	18.82	20.55	21.48	22.23	23.30
Kiribati	0	0	0	0	0	0
Korea, North	0	0	0	0	0	0

continues

TABLE 6.11 Continued
World Net Geothermal, Solar, Wind, and Wood and Waste Electric Power Consumption, 2000–2005 (Billion Kilowatt-hours)

Korea, South	0.40	0.34	0.77	0.27	0.41	0.42
Laos	0	0	0	0	0	0
Macau	0	0	0	0	0	0
Malaysia	0	0	0	0	0	0
Maldives	0	0	0	0	0	0
Mongolia	0	0	0	0	0	0
Nauru	0	0	0	0	0	0
Nepal	0	0	0	0	0	0
New Caledonia	0	0	0	0	0	0
New Zealand	4.05	3.85	4.04	3.87	3.58	3.99
Niue	0	0	0	0	0	0
Pakistan	0	0	0	0	0	0
Papua New Guinea	0	0	0	0	0	0
Philippines	11.05	9.92	9.73	9.33	9.77	9.41
Samoa	0	0	0	0	0	0
Singapore	0	0	0	0	0	0
Solomon Islands	0	0	0	0	0	0
Sri Lanka	0.003	0.004	0.004	0.003	0.003	0.002
Taiwan	0	0	0	0	0	0
Thailand	1.44	1.30	1.94	2.52	3.26	3.13
Tonga	0	0	0	0	0	0
U.S. Pacific Islands	0	0	0	0	0	0
Vanuatu	0	0	0	0	0	0
Vietnam	0	0	0	0	0	0
Wake Island	0	0	0	0	0	0
Asian and Oceanian totals	**46.28**	**47.63**	**51.44**	**53.07**	**56.42**	**59.34**
World totals	**242.63**	**253.02**	**284.45**	**308.16**	**341.45**	**369.71**

Source: Energy Information Administration.

TABLE 6.12
Energy Use by Non-OECD Economies by Region: Historical and Projected, 1990–2030
(Quadrillion Btu)

	Non-OECD Asia	Middle East	Africa	Central and South America	Non-OECD Europe and Eurasia	Total
1990	47.5	11.3	9.5	14.5	67.2	150.0
2000	73.1	17.3	12.0	20.9	44.9	168.2
2004	99.9	21.1	13.7	22.5	49.7	206.9
2010	131.0	26.3	16.9	27.7	54.7	256.7
2020	178.8	32.7	21.2	34.8	64.4	331.9
2030	227.6	38.2	24.9	41.4	71.5	403.5

Source: Energy Information Administration.

U.S. Data

TABLE 6.13
U.S. Energy Consumption by Energy Source, 2001–2005 [Quadrillion (10^{15}) Btu]

Energy Source	2001	2002	2003	2004	2005
Total	**96.563**	**98.101**	**98.450**	**100.586**	**100.942**
Fossil fuels	83.138	83.994	84.386	86.191	86.451
Coal	21.914	21.904	22.321	22.466	22.785
Natural gas	22.861	23.628	22.967	22.993	22.886
Petroleum	38.333	38.401	39.047	40.594	40.735
Electricity net imports	0.075	0.072	0.022	0.039	0.084
Nuclear electric power	8.033	8.143	7.959	8.222	8.160
Renewable energy	5.465	6.067	6.321	6.433	6.588
Conventional hydroelectric	2.242	2.689	2.825	2.690	2.703
Geothermal energy	0.311	0.328	0.331	0.341	0.343
Biomass	2.777	2.880	2.988	3.196	3.298
Solar energy	0.065	0.064	0.064	0.064	0.066
Wind energy	0.070	0.105	0.115	0.142	0.178

Source: Energy Information Administration.

TABLE 6.14

U.S. Individual State Energy Consumption Estimates by Source in 2004 [Trillion (10^{12}) Btu]

State	Total Energy	Coal	Natural Gas	Petroleum	Nuclear Electric Power	Hydroelectric Power	Biomass	Solar	Wind	Geothermal
Alabama	2,159.7	853.9	404.0	638.5	329.9	106.5	185.0	0	0	0
Alaska	779.1	14.1	411.8	334.8	0	15.0	3.3	0	0	0
Arizona	1436.6	425.4	354.9	562.8	293.1	69.9	8.7	(neg)	0	0
Arkansas	1135.9	270.2	228.9	388.3	161.1	36.5	76.0	0	0	0
California	8364.6	68.9	2474.2	3787.8	315.6	342.2	160.7	5.7	43.2	275.2
Colorado	1383.9	390.2	437.5	500.4	0	12.0	7.4	0	2.2	0
Connecticut	923.8	44.0	163.1	470.8	172.5	4.6	38.3	0	0	0
Delaware	304.8	53.6	49.9	140.8	0	0	1.3	0	0	0
Florida	4452.5	699.1	755.2	2119.9	325.5	2.7	176.1	0	0	0
Georgia	3141.1	835.0	410.3	1117.6	351.9	37.0	191.5	0	0	0
Hawaii	323.5	19.3	2.9	282.6	0	0.9	11.8	0	0.1	4.5
Idaho	499.8	12.3	77.2	156.6	0	84.8	25.8	0	0	0
Illinois	3960.5	1069.5	956.2	1374.5	959.8	1.5	48.5	0	0.8	0
Indiana	2945.7	1614.2	542.5	885.2	0	4.4	40.7	0	0	0

continues

TABLE 6.14 Continued
U.S. Individual State Energy Consumption Estimates by Source in 2004 [Trillion (10^12) Btu]

Iowa	1205.8	443.2	228.6	439.4	51.4	9.5	34.6	0	10.5	0
Kansas	1103.5	385.5	273.3	427.9	105.7	0.1	8.8	0	3.6	0
Kentucky	1956.4	961.8	231.8	727.8	0.0	37.9	26.7	0	0	0
Louisiana	3816.3	256.7	1400.0	1651.1	178.1	11.0	178.1	0	0	0
Maine	480.3	7.3	76.3	260.6	0.0	34.4	107.4	0	0	0
Maryland	1526.6	327.2	198.7	581.6	152.0	25.1	34.5	0	0	0
Massachusetts	1542.9	105.1	387.4	748.7	61.9	10.0	54.5	0	0	0
Michigan	3119.4	773.8	918.4	1033.9	318.7	15.4	90.6	0	(neg)	0
Minnesota	1826.3	378.8	363.3	714.9	138.6	7.4	57.2	0	8.1	0
Mississippi	1214.3	185.0	293.6	488.0	106.7	0.0	60.9	0	0	0
Missouri	1849.3	807.5	268.0	746.5	81.7	14.8	19.0	0	0	0
Montana	402.9	195.6	66.7	185.8	0.0	88.8	12.6	0	0	0
Nebraska	651.9	223.6	115.1	238.4	106.8	9.2	9.4	0	0.4	0
Nevada	693.7	193.6	219.5	264.3	0.0	16.2	3.4	0	0	27.3
New Hampshire	340.7	43.4	64.5	205.4	106.1	13.2	23.0	0	0	0
New Jersey	2630.2	112.7	647.1	1270.0	282.4	0.4	32.8	0	0	0
New Mexico	682.3	309.4	230.0	259.8	0.0	1.4	2.9	0	5.1	0

continues

TABLE 6.14 Continued
U.S. Individual State Energy Consumption Estimates by Source in 2004 [Trillion (10^{12}) Btu]

New York	4254	276.5	1119.9	1885.4	423.8	240.4	133.8	0	1.2	0
North Carolina	2715.6	782.7	232.7	992.9	418.0	54.5	86.3	0	0	0
North Dakota	402.3	398.4	60.3	133.8	0.0	15.5	3.4	0	2.1	0
Ohio	4022.8	1391.3	845.0	1368.3	166.3	7.3	42.9	0	0	0
Oklahoma	1484.9	372.1	555.9	532.5	0.0	29.8	27.9	0	5.7	0
Oregon	1093.6	36.5	243.2	390.8	0.0	331.5	46.4	0	6.2	0
Pennsylvania	4049.4	1474.3	732.5	1518.6	807.7	31.6	87.4	0	3.1	0
Rhode Island	226.4	0.1	74.6	99.3	0.0	0.1	3.8	0	0	0
South Carolina	1717.5	433.9	163.8	595.3	533.9	24.5	75.6	0	0	0
South Dakota	263.6	43.6	42.5	115.1	0.0	36.1	1.8	0	1.6	0
Tennessee	2,297.7	648.0	239.2	799.8	298.3	104.3	72.7	0	(neg)	0
Texas	11,971.40	1,626.0	3,941.2	5,801.3	421.6	13.0	75.3	0	31.4	0
Utah	740.2	399.7	164.9	278.8	0.0	4.5	4.1	0	0	4.1
Vermont	169.3	(neg)	8.7	95.1	40.2	11.9	10.0	0	0.1	0
Virginia	2,558.2	452.5	284.9	1,038.5	295.2	15.9	105.2	0	0	0
Washington	2,004.8	112.5	268.5	842.4	93.7	717.3	94.3	0	7.4	0
West Virginia	821.3	937.1	143.2	280.3	0.0	13.2	4.5	0	1.6	0

continues

TABLE 6.14 Continued

U.S. Individual State Energy Consumption Estimates by Source in 2004 [Trillion (10^{12}) Btu]

Wisconsin	1,847.7	499.2	384.9	630.7	124.0	19.9	74.1	0	1.0	0
Wyoming	454.4	500.5	111.6	158.8	0.0	5.9	0.9	0	6.2	0
U.S. totals	100,278.60	22,465.6	22,901.6	40,592.9	8,222.0	2,690.1	2682.8	5.8	141.7	311.3

(neg) = Value less than 0.5 trillion Btu.

Note: Totals may not equal sum of components due to equal rounding.

Source: Energy Information Administration.

Environmental Impacts

TABLE 6.15
World Carbon Dioxide Emissions from the Consumption and Flaring of Fossil Fuels, 2000–2005 (Million Metric Tons of Carbon Dioxide)

Region/Country	2000	2001	2002	2003	2004	2005
North America						
Bermuda	0.52	0.52	0.52	0.56	0.60	0.62
Canada	558.44	565.59	586.45	613.42	626.29	631.26
Greenland	0.56	0.57	0.58	0.58	0.59	0.59
Mexico	380.61	377.41	383.77	389.43	381.85	398.25
Saint Pierre and Miquelon	0.08	0.07	0.07	0.08	0.08	0.08
United States	5,823.48	5,723.12	5,763.14	5,812.81	5,935.47	5,956.98
North American totals	**6,763.68**	**6,667.27**	**6,734.52**	**6,816.88**	**6,944.88**	**6,987.78**
Central and South America						
Antarctica	0.23	0.23	0.23	0.24	0.24	0.24
Antigua and Barbuda	0.51	0.51	0.51	0.55	0.59	0.59
Argentina	137.15	127.46	119.50	133.34	139.79	146.64
Aruba	0.95	0.95	0.95	1.01	1.02	1.01
Bahamas, The	3.49	3.41	3.39	4.15	4.21	4.06
Barbados	1.70	1.64	1.64	1.68	1.15	1.44
Belize	0.66	0.86	0.80	0.84	0.90	0.93
Bolivia	9.23	8.02	9.30	11.63	11.65	11.96
Brazil	342.10	346.21	343.79	343.07	352.70	360.57
Cayman Islands	0.34	0.34	0.34	0.36	0.38	0.38
Chile	55.25	54.15	53.47	57.30	64.25	66.19
Colombia	57.73	58.10	52.73	56.73	55.44	58.80
Costa Rica	4.97	5.19	5.27	5.68	5.50	5.69
Cuba	32.67	33.04	33.77	32.95	32.78	32.98
Dominica	0.08	0.10	0.11	0.12	0.11	0.11
Dominican Republic	17.54	17.03	18.65	19.38	17.97	17.77
Ecuador	19.97	21.40	21.56	22.28	23.37	23.90
El Salvador	5.47	5.66	5.72	6.03	5.98	6.16
Falkland Islands	0.03	0.03	0.03	0.03	0.03	0.04
French Guiana	0.99	0.99	1.01	1.06	1.03	1.04

continues

TABLE 6.15 Continued
World Carbon Dioxide Emissions from the Consumption and Flaring of Fossil Fuels,
2000–2005 (Million Metric Tons of Carbon Dioxide)

Grenada	0.12	0.20	0.23	0.24	0.24	0.24
Guadeloupe	1.80	1.82	1.85	1.93	1.98	2.02
Guatemala	8.99	9.93	10.18	10.16	10.73	10.96
Guyana	1.67	1.72	1.74	1.69	1.52	1.58
Haiti	1.51	1.64	1.70	1.69	1.73	1.75
Honduras	4.54	5.27	5.91	5.95	6.93	7.13
Jamaica	10.74	10.81	11.06	11.49	11.52	11.55
Martinique	2.04	2.04	2.06	2.15	2.22	2.29
Montserrat	0.05	0.05	0.05	0.06	0.07	0.07
Netherlands Antilles	11.55	11.64	10.69	10.67	10.79	11.05
Nicaragua	3.67	3.91	4.02	4.08	4.28	4.30
Panama	12.82	13.33	12.56	12.49	14.02	14.33
Paraguay	3.53	3.42	3.57	3.70	3.72	3.85
Peru	26.74	25.95	25.98	26.30	28.84	31.31
Puerto Rico	27.57	34.34	34.93	37.16	37.88	39.02
Saint Kitts and Nevis	0.10	0.10	0.10	0.12	0.13	0.13
Saint Lucia	0.33	0.35	0.35	0.38	0.37	0.37
Saint Vincent/Grenadines	0.17	0.17	0.18	0.18	0.20	0.20
Suriname	1.60	1.68	1.77	1.82	1.90	1.86
Trinidad and Tobago	27.40	29.88	30.74	29.61	32.38	38.18
Turks and Caicos Islands	0.02	0.02	0.01	0.01	0.01	0.01
Uruguay	6.47	4.91	5.32	5.51	5.98	6.01
Venezuela	133.78	147.90	145.79	132.81	141.98	151.29
Virgin Islands, U.S.	9.79	14.21	13.65	15.17	18.05	16.05
Virgin Islands, British	0.06	0.06	0.06	0.07	0.09	0.09
Central and South American totals	**988.09**	**1,010.67**	**997.30**	**1,013.88**	**1,056.65**	**1,096.16**
Europe						
Albania	3.23	3.55	3.72	4.18	4.34	4.35
Austria	63.66	67.56	70.04	74.11	74.92	78.17
Belgium	144.12	142.24	138.48	145.49	144.03	135.81

continues

TABLE 6.15 Continued
World Carbon Dioxide Emissions from the Consumption and Flaring of Fossil Fuels,
2000–2005 (Million Metric Tons of Carbon Dioxide)

Bosnia and Herzegovina	13.89	14.29	14.80	15.39	16.04	17.45
Bulgaria	48.62	51.95	47.84	51.26	50.71	50.54
Croatia	19.96	20.10	21.05	22.38	21.65	21.46
Cyprus	7.48	8.14	8.05	8.05	8.53	8.81
Czech Republic	111.19	114.55	110.00	112.81	109.00	112.83
Denmark	54.10	55.64	52.26	61.21	55.90	50.96
Faroe Islands	0.67	0.67	0.67	0.67	0.68	0.68
Finland	50.54	54.02	54.35	66.53	61.50	52.25
Former Czechoslovakia	—	—	—	—	—	—
Former Serbia and Montenegro	42.60	43.34	48.49	52.00	53.30	52.56
Former Yugoslavia	—	—	—	—	—	—
France	398.97	402.52	398.60	405.15	412.49	415.27
Germany	847.99	868.64	841.34	865.22	867.29	844.17
Germany, East	—	—	—	—	—	—
Germany, West	—	—	—	—	—	—
Gibraltar	7.22	3.93	3.99	4.10	4.23	4.34
Greece	100.43	102.24	101.50	105.29	104.26	103.16
Hungary	55.59	57.16	57.03	59.19	57.91	59.84
Iceland	3.16	3.00	3.16	3.15	3.29	3.19
Ireland	40.38	42.98	42.24	40.79	42.83	44.10
Italy	444.45	441.60	448.46	470.31	465.55	466.64
Luxembourg	8.97	9.41	10.38	10.97	12.23	12.55
Macedonia	8.34	8.23	8.04	7.95	8.08	8.05
Malta	2.90	2.45	2.91	2.90	2.96	3.02
Montenegro	—	—	—	—	—	—
Netherlands	249.46	275.61	256.53	258.82	267.83	269.66
Norway	40.67	41.66	42.28	45.39	48.28	52.35
Poland	289.72	273.67	271.91	284.55	289.45	284.64
Portugal	63.07	60.66	65.45	61.38	62.89	64.97
Romania	92.57	101.24	99.28	99.01	98.82	99.34

continues

TABLE 6.15 Continued
World Carbon Dioxide Emissions from the Consumption and Flaring of Fossil Fuels,
2000–2005 (Million Metric Tons of Carbon Dioxide)

Serbia	—	—	—	—	—	—
Slovakia	37.06	38.58	37.91	39.06	38.02	37.81
Slovenia	15.56	15.96	16.40	16.67	16.68	16.77
Spain	323.13	327.82	344.87	353.25	373.67	387.11
Sweden	55.24	55.88	53.08	59.05	58.37	58.77
Switzerland	45.11	45.57	43.99	44.79	45.20	45.92
Turkey	200.31	182.73	193.00	204.43	208.84	230.04
United Kingdom	554.58	565.42	556.22	566.65	576.34	577.17
European totals	**4,444.93**	**4,503.01**	**4,468.32**	**4,622.16**	**4,666.10**	**4,674.75**
Eurasia						
Armenia	8.55	8.71	8.33	8.96	9.12	9.61
Azerbaijan	43.54	31.01	35.50	33.89	35.82	37.03
Belarus	59.70	53.49	50.45	55.35	54.49	61.41
Estonia	16.13	16.50	16.36	18.46	19.09	18.89
Former USSR	—	—	—	—	—	—
Georgia	4.60	4.02	4.53	3.76	4.14	4.72
Kazakhstan	135.26	146.07	149.68	159.14	168.73	198.01
Kyrgyzstan	7.16	6.48	5.74	5.17	5.81	5.28
Latvia	7.31	7.59	7.36	7.78	8.11	8.39
Lithuania	13.21	12.97	13.27	13.57	13.37	13.94
Moldova	5.85	5.79	6.30	7.02	6.61	7.16
Russia	1,580.21	1,569.51	1,569.42	1,627.41	1,668.69	1,696.00
Tajikistan	5.92	6.14	6.00	6.61	7.00	7.20
Turkmenistan	23.81	29.60	33.98	43.76	46.72	49.64
Ukraine	322.51	314.50	322.29	352.64	342.08	342.57
Uzbekistan	105.78	110.66	113.40	114.11	120.93	117.97
Eurasian totals	**2,339.54**	**2,323.04**	**2,342.64**	**2,457.64**	**2,510.72**	**2,577.82**
Middle East						
Bahrain	20.18	20.70	21.62	22.35	23.05	25.18
Iran	318.58	332.01	362.48	384.33	403.91	450.68

continues

TABLE 6.15 Continued
**World Carbon Dioxide Emissions from the Consumption and Flaring of Fossil Fuels,
2000–2005 (Million Metric Tons of Carbon Dioxide)**

Iraq	73.14	77.18	78.91	70.99	77.80	98.13
Israel	62.24	66.55	68.59	64.11	65.51	65.01
Jordan	15.51	15.18	15.87	16.95	18.09	18.67
Kuwait	59.21	60.08	55.83	63.02	67.39	76.69
Lebanon	16.32	15.85	15.88	16.01	15.62	16.17
Oman	21.68	22.09	22.87	22.35	24.22	29.72
Qatar	34.54	27.44	29.13	32.35	38.48	53.45
Saudi Arabia	289.33	299.89	309.62	344.78	385.76	412.35
Syria	51.40	48.99	51.44	51.30	51.22	49.78
United Arab Emirates	109.65	118.13	125.55	126.38	132.76	137.82
Yemen	9.42	9.64	9.83	16.53	16.52	17.15
Middle Eastern totals	**1,081.19**	**1,113.73**	**1,167.63**	**1,231.45**	**1,320.32**	**1,450.81**
Africa						
Algeria	83.26	78.78	81.25	81.27	79.70	88.10
Angola	13.01	14.97	15.61	18.85	19.77	20.39
Benin	1.64	1.64	1.73	2.04	2.38	2.27
Botswana	4.33	3.74	3.82	3.48	3.96	3.92
Burkina Faso	1.05	1.06	1.09	1.13	1.15	1.17
Burundi	0.37	0.39	0.40	0.41	0.38	0.41
Cameroon	6.81	6.45	6.16	6.10	6.44	6.81
Cape Verde	0.17	0.14	0.14	0.14	0.29	0.28
Central African Republic	0.32	0.33	0.34	0.34	0.32	0.34
Chad	0.19	0.19	0.20	0.20	0.19	0.19
Comoros	0.10	0.10	0.10	0.10	0.11	0.10
Congo (Brazzaville)	3.01	3.20	2.98	3.06	3.07	5.31
Congo (Kinshasa)	2.69	1.99	1.80	1.84	2.37	2.37
Cote d'Ivoire (Ivory Coast)	7.32	5.83	5.88	5.87	6.40	6.42
Djibouti	1.84	1.86	1.86	1.90	1.93	1.95
Egypt	119.04	131.67	132.96	142.46	152.00	161.79
Equatorial Guinea	2.05	2.05	3.92	3.47	3.81	4.87

continues

TABLE 6.15 Continued
World Carbon Dioxide Emissions from the Consumption and Flaring of Fossil Fuels,
2000–2005 (Million Metric Tons of Carbon Dioxide)

Eritrea	0.64	0.75	0.71	0.80	0.76	0.78
Ethiopia	3.43	3.52	3.86	4.15	4.31	4.37
Gabon	5.06	4.94	4.69	4.88	4.91	4.95
Gambia, The	0.26	0.28	0.29	0.29	0.29	0.30
Ghana	5.28	5.14	5.41	5.90	6.42	6.67
Guinea	1.29	1.30	1.31	1.32	1.33	1.34
Guinea-Bissau	0.35	0.36	0.37	0.38	0.38	0.38
Kenya	8.65	7.93	7.89	8.31	9.16	9.88
Lesotho	0.21	0.21	0.21	0.21	0.21	0.21
Liberia	0.44	0.46	0.50	0.51	0.52	0.53
Libya	41.63	43.19	48.02	46.04	51.92	53.47
Madagascar	1.80	1.97	2.28	2.36	2.60	2.54
Malawi	0.75	0.83	0.82	0.76	0.89	0.86
Mali	0.56	0.56	0.62	0.62	0.65	0.66
Mauritania	3.23	3.16	3.05	3.14	2.63	2.63
Mauritius	3.47	4.03	3.58	3.69	3.93	4.01
Morocco	30.85	33.46	33.40	34.15	35.15	38.89
Mozambique	1.27	1.49	1.68	1.85	2.12	2.30
Namibia	1.80	2.16	2.21	2.41	2.55	2.67
Niger	1.16	1.17	1.21	1.23	1.23	1.23
Nigeria	80.42	91.16	90.52	92.37	91.83	105.19
Reunion	2.50	2.60	2.58	2.58	2.63	2.65
Rwanda	0.76	0.77	0.77	0.78	0.76	0.78
Saint Helena	0.02	0.01	0.01	0.01	0.01	0.01
Sao Tome and Principe	0.09	0.09	0.09	0.09	0.09	0.10
Senegal	4.40	4.38	4.43	4.47	5.27	5.49
Seychelles	0.58	0.60	0.88	0.86	0.89	0.92
Sierra Leone	0.91	0.93	0.95	0.97	1.15	1.18
Somalia	0.72	0.72	0.72	0.72	0.72	0.75
South Africa	383.42	390.74	377.11	409.38	438.13	423.81

continues

TABLE 6.15 Continued
World Carbon Dioxide Emissions from the Consumption and Flaring of Fossil Fuels,
2000–2005 (Million Metric Tons of Carbon Dioxide)

Sudan	6.47	7.42	8.59	9.72	10.67	10.79
Swaziland	1.26	1.16	1.13	1.15	1.15	1.14
Tanzania	2.70	3.06	3.47	3.64	3.91	3.97
Togo	1.40	1.00	1.28	2.14	2.25	2.38
Tunisia	19.56	21.30	21.03	20.68	20.98	22.24
Uganda	1.26	1.38	1.45	1.58	1.60	1.62
Western Sahara	0.26	0.27	0.27	0.27	0.27	0.28
Zambia	1.88	2.05	2.11	2.22	2.32	2.45
Zimbabwe	13.28	12.73	12.16	10.96	9.58	11.78
African totals	**881.24**	**913.67**	**911.89**	**960.27**	**1,010.45**	**1,042.92**
Asia and Oceania						
Afghanistan	1.33	0.98	1.00	0.80	0.86	0.98
American Samoa	0.56	0.56	0.57	0.57	0.57	0.60
Australia	352.57	366.75	374.60	375.27	381.22	406.64
Bangladesh	29.20	32.34	34.00	36.02	37.97	39.82
Bhutan	0.29	0.30	0.30	0.30	0.31	0.32
Brunei	3.78	4.20	4.83	5.15	5.75	6.44
Burma	8.94	8.66	9.21	10.18	11.15	13.87
Cambodia	0.54	0.55	0.55	0.57	0.55	0.57
China	2,912.59	3,050.88	3,376.15	3,983.12	4,753.33	5,322.69
Cook Islands	0.06	0.06	0.06	0.06	0.06	0.07
East Timor	—	—	—	NA	NA	NA
Fiji	0.83	0.82	1.19	1.50	1.28	1.34
French Polynesia	0.67	0.67	0.69	0.85	0.85	0.87
Guam	2.86	3.05	2.09	2.43	1.95	2.18
Hawaiian Trade Zone	—	—	—	—	—	—
Hong Kong	55.37	59.87	65.62	69.53	77.44	75.06
India	994.07	1,016.67	1,014.71	1,029.08	1,128.92	1,165.72
Indonesia	271.05	296.84	311.39	314.95	341.56	359.47
Japan	1,190.16	1,177.70	1,185.79	1,234.15	1,241.91	1,230.36

continues

TABLE 6.15 Continued
World Carbon Dioxide Emissions from the Consumption and Flaring of Fossil Fuels,
2000–2005 (Million Metric Tons of Carbon Dioxide)

Kiribati	0.03	0.03	0.03	0.03	0.03	0.03
Korea, North	68.73	70.41	67.27	68.43	69.83	73.50
Korea, South	439.98	446.13	465.24	479.40	488.14	499.63
Laos	0.88	1.01	1.03	1.05	1.05	1.06
Macau	1.57	1.65	1.83	1.89	2.31	2.35
Malaysia	111.31	124.16	137.66	148.68	164.43	155.51
Maldives	0.48	0.51	0.94	1.06	0.74	0.76
Mongolia	6.60	7.02	7.46	7.02	7.64	7.87
Nauru	0.16	0.16	0.17	0.17	0.17	0.18
Nepal	3.07	3.22	2.63	2.77	2.87	2.95
New Caledonia	1.98	1.99	2.27	2.36	2.52	2.51
New Zealand	34.87	37.45	37.90	39.53	37.26	37.82
Niue	0.003	0.003	0.003	0.003	0.003	0.003
Pakistan	108.20	104.47	105.48	107.60	113.45	121.49
Papua New Guinea	2.58	2.58	3.66	4.23	4.28	4.35
Philippines	70.31	71.05	71.62	71.87	74.21	78.06
Samoa	0.14	0.14	0.15	0.15	0.15	0.16
Singapore	106.81	107.60	109.37	111.54	125.40	133.88
Solomon Islands	0.18	0.18	0.18	0.18	0.19	0.19
Sri Lanka	11.25	10.99	11.36	11.63	11.94	12.38
Taiwan	248.54	245.22	269.43	285.65	281.87	284.40
Thailand	160.59	170.88	185.69	204.18	223.73	234.16
Tonga	0.13	0.12	0.11	0.11	0.12	0.13
U.S. Pacific Islands	0.30	0.30	0.30	0.30	0.30	0.29
Vanuatu	0.08	0.08	0.09	0.09	0.09	0.09
Vietnam	47.39	50.57	57.69	61.77	76.97	80.38
Wake Island	1.32	1.32	1.34	1.34	1.34	1.36
Asian and Oceanian totals	**7,252.33**	**7,480.15**	**7,923.64**	**8,677.55**	**9,676.73**	**10,362.49**
World totals	**23,751.01**	**24,011.54**	**24,545.94**	**25,779.84**	**27,185.85**	**28,192.74**

Source: Energy Information Administration.

Reference

Energy Information Administration. *Annual Energy Review 2006, 2007.* Report No. DOE/EIA-0384, June 27, 2006.

7

Directory of Organizations

Nongovernmental Organizations

American Council on Renewable Energy (ACORE)
P.O. Box 33518
Washington, DC 20033-3518
(202) 393-0001
Fax: (202) 393-0606
Web site: www.acore.org

The American Council on Renewable Energy (ACORE) is a non-profit organization based in Washington, D.C. It was founded in November 2001, with a focus on promoting a transition from fossil fuels to renewable energy, including solar, wind, geothermal, hydro and ocean, waste and energy fuels, and biomass energy and biofuels. To do this, ACORE has a three-prong strategy. It convenes meetings, conferences, roundtables, forums, and other gatherings that help to inform people and provide an opportunity for them to connect with others on renewable energy issues related to trade, finance, and policy. The second part of its strategy is information publishing, including the annual report "Renewable Energy in America." This report provides an updated assessment of the status and future development of various renewable energy technologies. The third part is communications; although this involves the education of the public, the media, and policy makers, the organization specifically says it is not a lobbying organization. Its intent is to provide information about renewable energy that will better inform, not necessarily persuade, decision makers.

Environmental and Energy Study Institute (EESI)
122 C Street NW, Suite 630
Washington, DC 20001
(202) 628-1400
Fax: (202) 628-1825
E-mail: eesi@eesi.org
Web site: www.eesi.org

The Environmental and Energy Study Institute (EESI) is a non-profit organization started in 1984 by a bipartisan group of members of the U.S. Congress who were concerned about energy and environmental issues. The organization promotes environmentally sustainable societies through the production of information and the formulation of public policy initiatives. The goal of these initiatives is to promote the transition to the social and economic conditions that will facilitate sustainable living conditions. EESI provides education to policy makers and policy analysis projects in the areas of energy efficiency and renewable energy, global climate change, agriculture, biofuels, smart growth, and clean bus technologies.

Geothermal Resources Council (GRC)
P.O. Box 1350
2001 Second Street, Suite 5
Davis, CA 95617
(530) 758-2360
Fax: (530) 758-2839
E-mail: grc@geothermal.org
Web site: www.geothermal.org

The Geothermal Resources Council (GRC) is a nonprofit educational association that was formed in 1970. It has members in more than 20 countries. The mission of the council is to become the premier geothermal energy educational association throughout the world. The council seeks to promote the development of geothermal resources across the globe by collecting and distributing data and technological information. It also encourages research into environmentally sustainable exploration and development of geothermal energy sources. To do this, it cooperates and communicates with academic institutions, the geothermal industry, and the government on development and use of geothermal resources with a concern for economical and environmentally sound solutions. It also provides a forum for members of the international geothermal community to transfer objective information about geothermal energy and its development.

Global Environmental Facility (GEF)
1818 H Street NW
Washington, DC 20433
(202) 473-0508
Fax: (202) 522-3240/-3245
E-mail: gef@gefweb.org
Web site: www.gefweb.org
Contact: GEF Secretariat

Global Environmental Facility (GEF) is an independent financial organization whose members consist of both developing and developed countries. Founded in 1991, it began funding projects in developing countries with a focus on improving the global environment and sustainability of the local communities. GEF grants fund projects that address global environmental problems like biodiversity, climate change, international waters, land degradation, the ozone layer, and persistent organic pollutants (POPs). The organization works in cooperation with various United Nations agencies, the World Bank, the European Bank, and similar regional organizations. Its connection to renewable energy is due to its work on climate change. In an effort to reduce worldwide emissions of greenhouse gas emissions, GEF funds renewable energy, energy efficiency, and sustainable transportation projects in the developing world. It is the largest single source of funding for the environment.

Green Power Market Development Group
10 G Street NE, Suite 800
Washington, DC 20002
(202) 729-7624
E-mail: chanson@wri.org.
Web site: www.thegreenpowergroup.org
Contact: Craig Hanson

The Green Power Market Development Group was convened in 2000 by the World Resources Institute, an environmental think tank, and consists of representatives from 12 leading corporations, including companies like British Petroleum, General Motors, and Monsanto, and other "green power affiliates." The group is a collaboration among these members in an effort to increase the shares of green power by 1,000 megawatts by 2010. Its strategy for doing this is to create economies of scale for technologies and reduce market barriers by providing information for customers. Through the purchase of renewable energy credits and the development of projects that use energy from wind, geothermal, and hydrogen fuel cells, the group generated 360 megawatts of power between 2001 and 2005.

Interstate Renewable Energy Council (IREC)
P.O. Box 1156
Latham, NY 12110-1156
(518) 458-6059
E-mail: info@irecusa.org
Web site: www.irecusa.org/

The Interstate Renewable Energy Council (IREC) is a collaborative effort among state and local offices with national laboratories and solar and renewable energy organizations and companies that began in 1982. Its purpose is to expedite the transition to renewable energy through the actions of state and local governments and communities that support market-oriented approaches and that facilitate the implementation of uniform standards within the industry. It focuses particularly on shaping and working with administrative rules and regulations, getting small-scale projects connected to the grid, building the credibility and competency in renewable energy professionals, and providing information. It also promotes the Solar Rating and Certification Corporation, which aims to develop and implement standardized rating and certification of solar energy technologies. It manages the Database of State Incentives for Renewables and Efficiency (DSIRE), a comprehensive and helpful database available to the public.

Public Renewables Partnership
Lawrence Berkeley National Laboratory
University of California-Berkeley
(510) 486-5229
E-mail: whgolove@lbl.gov
Web site: www.repartners.org
Contact: Bill Golove, national chair

Public Renewables Partnership serves to inform the utility decision makers for public organizations, cooperatives, and Native American tribes about their renewable energy options and prospects. To do this, it evaluates renewable energy resource potential, analyzes the market and energy portfolios, provides assistance with strategic planning, and maintains a Web site with a host of information. Its underlying aim is to increase the utilization of renewable energy technologies by providing the information needed to make the decision. This means that it provides an efficient means for decision makers to access the information needed to make the best decision they can regarding their utility use or production. It does this by maintaining an up-to-date database of news in the field, finding funding for and conducting research projects related to renewable energy, and serving as a bridge between various stakeholders to coordinate collaboration among them.

Sustainable Energy Coalition (SEC)
1612 K Street NW, Suite 202-A
Washington, DC 20006-2802
Web site: www.sustainableenergycoalition.org/

The Sustainable Energy Coalition (SEC) is comprised of more than 60 national and state-level business, environmental, consumer, and energy policy organizations. It was founded in 1992 to promote increased federal support for energy efficiency and renewable energy technologies. It also works to reduce federal support for unsustainable energy developments. The SEC works to influence policy decisions regarding the federal budget, electric utility restructuring, pollution prevention, climate change, and tax policies. SEC hosts the annual Renewable Energy and Energy Efficiency Expo in collaboration with U.S. House of Representatives and Senate renewable energy and energy efficiency caucuses, which provides a forum for leading businesses and experts to present viable technologies and analyze their potential to offset climate change.

Research Institutes

National Center for Photovoltaics
901 D Street SW, Suite 930
Washington, DC 20024-2157
(303) 384-6491
E-mail: david_christensen@nrel.gov

Web site: www.nrel.gov/pv/
Contact: David Christensen

The National Renewable Energy Laboratory's (NREL) National Center for Photovoltaics (PV) research is focused on decreasing the nation's reliance on fossil fuel–generated electricity by lowering the cost of delivered electricity and improving the efficiency of PV modules and systems. NREL's PV research contributes to these goals through fundamental research, advanced materials and devices, and technology development. NREL's capabilities include research and development, testing and evaluation, and deployment. Its PV R&D (research and development) emphasizes innovative research, thin film development, manufacturing R&D, and systems development and reliability. It conducts this research in collaboration with universities and solar industry through research partnerships and direct support of the NREL Solar Program and the Solar America Initiative.

National Renewable Energy Laboratory (NREL)
1617 Cole Boulevard
Golden, CO 80401-3393
(303) 275-3000
E-mail: info@nrel.gov
Web site: www.nrel.gov

The National Renewable Energy Laboratory (NREL) is the nation's primary laboratory for renewable energy and energy efficiency R&D. NREL's mission and strategy are focused on advancing the U.S. Department of Energy's and U.S. energy goals. The laboratory's scientists and researchers support critical market objectives to accelerate research from scientific innovations to market-viable alternative energy solutions. At the core of this strategic direction are NREL's research and technology development areas. These areas span from understanding renewable resources for energy, to the conversion of these resources to renewable electricity and fuels, and ultimately to the use of renewable electricity and fuels in homes, commercial buildings, and vehicles. The laboratory thereby directly contributes to the nation's goal of finding new renewable ways to power homes, businesses, and cars.

A critical part of the Lab's mission is the transfer of NREL-developed technologies to renewable energy markets. NREL's Technology Transfer Office supports laboratory scientists and engineers in the successful and practical application of their expertise and the technologies they develop. NREL's world-class R&D staff and facilities are recognized and valued by industry, as demonstrated through hundreds of collaborative research projects and licensed technologies with public and private partners. NREL's innovative technologies have also been recognized with 39 R&D 100 awards—the most per staff member of any DOE laboratory. The engineering and science behind these technology transfer successes and awards demonstrates NREL's commitment to developing and applying innovative renewable energy solutions for the nation's secure and sustainable energy future. NREL PV R&D is performed under the National Center for Photovoltaics.

National Wind Technology Center (NWTC)
1617 Cole Boulevard
Golden, CO 80401-3393
(303) 384-6945
Web site: www.nrel.gov/wind/nwtc.html

Much of the wind industry's success can be attributed to the research conducted at NREL's National Wind Technology Center (NWTC). Funded by the U.S. Department of Energy Wind Energy Technologies Program, research conducted at the NWTC has led to the development of multimegawatt wind turbines that produce electricity at a cost that is starting to compete with conventional energy sources in the marketplace. To make wind energy fully cost competitive and to increase wind energy development, researchers at the NWTC are working in partnership with industry to develop larger, more efficient, utility-scale wind turbines for land-based and offshore installations, as well as more efficient, quiet, small wind turbines for distributed applications.

Renewable Energy Policy Project (REPP)
1612 K Street NW, Suite 202
Washington, DC 20006
(202) 293-2898
Fax: (202) 293-5857
E-mail: info2@repp.org
Web site: www.crest.org

The Renewable Energy Policy Project (REPP) was started in 1995 with the support of the Energy Foundation and the U.S. Department of Energy. Its main task is to promote the advancement of renewable energy through policy research. REPP research focuses on the relationships among policy, markets, and consumer demand. The information is disseminated by the organization to policy makers, renewable energy companies, environmental advocates, and the public. In 1999, REPP joined with the Center for Renewable Energy and Sustainable Technology (CREST) to further public involvement in the process through online forums and discussion groups and interactive software. The project also provides reports for states containing information about the status of renewable energy, its potential, and the policy options.

Sandia National Laboratories
P.O. Box 5800
Albuquerque, NM 87185
(505) 284-5200
E-mail: info@sandia.gov
Web site: www.sandia.gov

Started in 1949, Sandia National Laboratories is a government-owned and contractor-operated (GOCO) facility. Sandia Corporation, a Lockheed Martin company, manages Sandia for the U.S. Department of Energy's National Nuclear Security Administration. It develops science-based technologies that support goals associated with national security. Through science and technology, people, infrastructure, and partnerships, Sandia's mission is to meet national needs in five key areas: ensuring that the nuclear weapons stockpile is safe, is secure, is reliable, and can support the United States' deterrence policy; enhancing the surety of energy and other critical infrastructures; reducing the proliferation of weapons of mass destruction, the threat of nuclear accidents, and the potential for damage to the environment; assessing defense systems and addressing new threats to national security; and helping to ensure homeland security from terrorist threats. The Sandia National Laboratories Corporate Archives and History Program exist to preserve, document, and disseminate Sandia's history. As a critical link in the nuclear weapons development chain, Sandia played a large role in the history of the Cold War, while its reimbursable projects have responded to U.S. national security policy more broadly. To preserve the crucial records of Sandia's history, the Corporate Archives collects and preserves information from Sandia's past, and the History Program brings Sandia's story to a broader public.

University of Delaware Center for Energy and Environmental Policy
Newark, DE 19716
(302) 831-8405
Fax: (302) 831-3098
E-mail: jbbyrne@udel.edu
Web site: ceep.udel.edu/ceep.html

The University of Delaware is a leading institution for research into renewable energy and renewable energy policy. The university has several organizations within it dedicated to the advancement of renewable energy technologies. The Center for Energy and Environmental Policy was established in 1980 after the oil shocks of the 1970s highlighted the need for concentrated research in energy and the environment. The Center is the leading institution for interdisciplinary education, research, and advocacy in energy and environmental policy. The research and advocacy programs of the Center address issues at the international, regional, national, state, and local levels. The university also houses the Institute of Energy Conversion, which was created in 1972 as the world's first laboratory dedicated to the research and development of thin-film photovoltaic and solar thermal systems. The Institute leads the Consortium for Very High Efficiency Solar Cells (VHESC), which has achieved record levels of efficiency in solar cell technologies. Most recently, VHESC achieved a 42.8 percent efficiency solar cell.

University of Massachusetts–Amherst Center for Energy Efficiency and Renewable Energy (CEERE)
College of Engineering
160 Governor's Drive
Amherst, MA 01003
(413) 545-4359
E-mail: rerl@ecs.umass.edu
Web site: www.ceere.org/
Contact: Dr. James Manwell

The University of Massachusetts houses the Center for Energy Efficiency and Renewable Energy (CEERE), which seeks to provide technological and economic solutions to environmental problems resulting from energy production, industrial, manufacturing, commercial activities, and land use practices. CEERE partners with state, federal, and private organizations to conduct research in one of four areas: renewable energy resources, building energy efficiency, industrial energy efficiency, and environmental technology. The Massachusetts Energy Efficiency Partnership is a program that coordinates the major gas and electric utilities in the state to conserve energy in the commercial and industrial sectors.

Associations

American Wind Energy Association (AWEA)
1101 14th Street NW, 12th Floor
Washington, DC 20005
(202) 383-2500
Fax: (202) 383-2505
E-mail: windmail@awea.org
Web site: www.awea.org/

The American Wind Energy Association (AWEA) is one of the biggest trade associations related to wind power in the United States. It promotes wind power development through advocacy, community, and education. It is well-known for providing up-to-date, reliable, and accurate information about the wind power issues in the United States and the world. This information includes data and statistics about domestic and international market development. Its annual conference, WINDPOWER, is the largest in North America, attended by many of the 1,000 members of the association. It also publishes the only weekly newsletter about wind energy. The Association interacts with members of Congress to ensure that wind energy interests are represented in renewable energy legislation.

Biomass Energy Research Association (BERA)
901 D Street SW, Suite 100
Washington, DC 20024
(847) 381-6320
Fax: (847) 382-5595
E-mail: beraorg2@excite.com
Web site: www.bera1.org

The Biomass Energy Research Association (BERA) is a nonprofit organization that was started in 1982 by researchers from the biomass industry and academic institutions. Its purpose is to develop and commercialize biomass energy technology for the public good. To do this, BERA encourages research in the private and public sectors, facilitates the exchange of information between them, and educates the public on the potential of biomass. The association also monitors federal biomass legislation and contributes its knowledge to discussions about research funding needs by participating in public hearings and issuing an annual position statement on federal funding of research on bioenergy. The association also publishes a catalog of products being sold by its members for the purpose of promoting U.S. exports of these goods.

British Wind Energy Association (BWEA)
Renewable Energy House
1 Aztec Row, Berners Road
London N1 0PW, United Kingdom
020 7689 1960
Fax: 020 7689 1969
E-mail: info@bwea.com
Web site: www.bwea.com

The British Wind Energy Association (BWEA) is the trade and professional body for the United Kingdom wind and marine renewables industries. Formed in 1978, and with over 310 corporate members, BWEA is the leading renewable energy trade association in the U.K. In 2004, BWEA expanded its mission to champion wave and tidal energy and use the association's experience to guide these technologies along the same path to commercialization.

Danish Wind Industry Association
Vester Voldgade
106 DK-1552
Copenhagen V, Denmark
45 3373 0330
Fax: 45 3373 0333
E-mail: danish@windpower.org
Web site: www.windpower.org/en/core.htm/

Danish Wind Industry is a nonprofit association that promotes the domestic and international adoption of wind energy technologies. Founded in 1981, the Danish Wind Industry Association now represents virtually all of Danish wind turbine manufacturers with energy measured in megawatts. It also represents more than 140 companies associated with wind energy activities in Denmark. Using a variety of methods to promote energy, the Danish Wind Industry Association publishes information about wind energy online through its Web site and in printed form, including an annual report of the state of the industry. Additionally, it engages in political advocacy for policies that support wind energy development in Denmark. The Association is also a member of the European Wind Energy Association.

European Renewable Energy Centres Agency (EUREC)
Renewable Energy House
63-65 rue d'Arlon
B-1040 Brussels, Belgium
32 2 546 1930
Fax: 32 2 546 1934
E-mail: info@eurec.be
Web site: www.eurec.be/

The European Renewable Energy Centres Agency (EUREC) was established in 1991 to strengthen European research, demonstration, and development efforts in all renewable energy technologies. EUREC is an independent member-based association with 48 prominent research groups from all over Europe. It is the voice of renewable energy research in Europe. Its members conduct research in the fields of solar building design, wind, photovoltaics, biomass, small hydro, solar thermal power stations, ocean energy, solar chemistry and solar materials, hybrid systems, renewable energy in developing countries, and integration of renewable energy in the energy infrastructure. EUREC is also a founding member of the European Renewable Energy Council.

European Biomass Association (AEBIOM)
Croix du Sud
2 bte 11 Louvain-la-Neuve
1348 Brussels, Belgium
32 10 47 3455
Fax: 32 10 47 3455
E-mail : jossart@aebiom.org
Web site: www.aebiom.org
Contact: Jossart Jean-Marc, secretary-general

The European Biomass Association (AEBIOM) is a nonprofit, international organization founded in 1990. Based in Brussels, the AEBIOM serves as an umbrella organization for 28 national associations across Europe. Its mission is to represent bioenergy at the European Union (EU) level. The associations act as members in the General Assembly, which, along with the Steering Committee and the board, manages AEBIOM. The activities of AEBIOM range from project management, EU affairs, and event organization to information dissemination and communication using tools such as a bimonthly newsletter; a yearly journal, *Biomass News;* and other publications. To encourage the production and use of biomass throughout Europe, AEBIOM undertakes a number of strategies. AEBIOM develops and advertises global solutions to encourage the production of biomass and undertakes studies toward the same end. It coordinates among the different international associations and encourages information sharing among members. It provides assistance to those interested in setting up new national associations. It cooperates with important European institutions such as the European Commission, European Parliament, and Economic and Social Council. AEBIOM organizes seminars, conferences, and information and awareness campaigns. AEBIOM also promotes technology transfer to developing countries.

European Geothermal Energy Council (EGEC)

63-65, rue d'Arlon
B-1040 Brussels, Belgium
32 2 400 1024
Fax: 32 2 400 1010
E-mail: info@egec.org
Web site: www.egec.org/

Formed in 2003, the European Geothermal Energy Council (EGEC) is an international association that promotes use of geothermal energy in Europe and the export of geothermal technology from the European geothermal industry to the rest of the world. Its main priority is to encourage research and development of geothermal resources in Europe and to make the results of this research useful for the public in exploiting geothermal resources. The EGEC works with European institutions to encourage the adoption of legislation and regulations that will help geothermal compete with conventional energy sources. The organization also works to have the environmental benefits considered when decisions about fiscal support are made. The EGEC cooperates with national geothermal associations, including the International Geothermal Association (IGA) and its European branch, as well as with any other organizations promoting research on and application of renewable energy sources. It disseminates information through publications, meetings, and discussions at the European level.

European Ocean Energy Association
63-65 rue d'Arlon
B-1040 Brussels, Belgium
32 0 2 400 1040
Fax: 32 0 2 791 9000
Web site: www.eu-oea.com/

The European Ocean Energy Association is a nonprofit organization with a member-elected board of directors created to represent all actors in the ocean energy sector at a European level. Because ocean energy is still small compared to other types of renewable energy, most of the members are just getting businesses started. To make ocean energy a more viable option, the European Ocean Energy Association seeks to make the source more visible and promote its use on an international level. The association provides a forum for the burgeoning industry to explore avenues for development of ocean energy. It belongs to the European Renewable Energy Council, which has access to decision makers and governmental officials at the EU level.

European Photovoltaic Industry Association (EPIA)
63-65 rue d'Arlon
B-1040 Brussels, Belgium
32 2 465 3884
Fax: 32 2 400 1010
E-mail: com@epia.org
Web site: www.epia.org

Founded in 2006, the European Photovoltaic Industry Association (EPIA) is the world's largest solar electricity industry association. EPIA's membership is comprised of over 100 companies from over 20 countries. The Association promotes the development of photovoltaic markets at the national, European, and global levels. EPIA represents its members when collaborating with European institutions like the European Commission. It provides policy analysis to its members to predict what effect policies and programs that are passed will have on the industry. EPIA also advocates for specific legislation to encourage photovoltaic market development. It also coordinate legislation with industrial advances. Like many of the other associations mentioned in this chapter, EPIA facilitates interaction among its members and supports national organizations in achieving their local objectives. EPIA works to organize conferences in areas where solar electricity has strong potential but requires additional support. EPIA is also a founding member of the Alliance for Rural Electrification (ARE), which promotes the use of renewable energies for rural electrification.

European Renewable Energies Federation (EREF)
73 avenue de la Fauconnerie
1170 Brussels, Belgium
32 2 672 4367
Fax: 32 2 672 7016
E-mail: info@eref-europe.org
Web site: www.eref-europe.eu

The European Renewable Energies Federation (EREF) is the European interest and lobby group for independent producers of electricity and fuels from renewable energy sources like wind, solar, small-scale hydro, and biomass. The members include national associations of renewable energy producers and supporting agencies from many EU member states. EREF members have a combined total of approximately 17.000 megawatts of installed capacity in renewable energy electricity in the European Union. EREF is the voice of renewable energy producers. It acts as a watchdog of the energy sector and provides information to the public and the media to promote renewable energy and regional sustainable development.

European Small Hydropower Association (ESHA)
Renewable Energy House
63-65 rue d'Arlon
B-1040 Brussels, Belgium
32 2 546 1945
Fax: 32 2 546 1947
E-mail: info@esha.be
Web site: www.esha.be

The European Small Hydropower Association (ESHA) is a non-profit international association that represents national associations and individuals who belong to the small hydropower sector. ESHA was founded in 1989 as an initiative of the European Commission. ESHA is one of the founding members of the European Renewable Energy Council, which groups all main European renewable energy industry and research associations. ESHA is located in the Renewable Energy House, in Brussels, with other EREC members. ESHA believes that the benefits of small hydropower can be achieved only if policies favorable for its development are sought at the European, national, and local levels. Like many of the other associations that belong to EREC, ESHA encourages the sharing of technology and information among its members. It also engages in studies that analyze the various factors that affect the development and adoption of small hydropower.

European Wind Energy Association (EWEA)
63-65 rue d'Arlon
B-1040 Brussels, Belgium
32 2 546 1940
Fax: 32 2 546 1944
E-mail: ewea@ewea.org
Web site: www.ewea.org/

European Wind Energy Association (EWEA) is considered to be one of the most powerful wind energy networks in the world. With members from 40 countries including over 300 companies, associations, and research institutions, EWEA's members include manufacturers making up 98 percent of the world wind power market, component suppliers, research institutes, national wind and renewables associations, developers, electricity providers, finance and insurance companies, and consultants. With such a broad array of members, the EWEA has become the most influential organization in the world wind energy network. The EWEA secretariat is located at the Renewable Energy House in Brussels. The secretariat works to coordinate EU wind policy, communications, research, and analysis. It serves as a manager of various European wind energy projects, hosts events, and supports the needs of its members. EWEA is a founding member of the European Renewable Energy Council, which serves as an umbrella organization for the eight major renewable industry and research associations. It is also the founder of the Global Wind Energy Council

Geothermal Energy Association
209 Pennsylvania Avenue SE
Washington, DC 20003
(202) 454-5261
Fax: (202) 454-5265
E-Mail: daniela@geo-energy.org
Web site: www.geo-energy.org

The Geothermal Energy Association is a trade association that consists of companies that support the expansion of geothermal energy. To encourage the adoption of geothermal energy, the Association advocates favorable policies, provides a forum for discussion of geothermal issues, encourages research and development of geothermal energy resources, gathers data and research, represents the industry before the government in public hearings, and provides education and outreach to the public. Specifically, the Association is advocating the adoption of a geothermal energy target of 20 percent of the total power shares. The association also hosts a trade show to facilitate the dissemination of technology and information.

German Wind Energy Association (BWE)
Herrenteichsstr. 1
49074 Osnabrück, Germany
49 0 541 350 60 0
Fax: 49 0 541 350 60 30
E-mail: info@wind-energie.de
Web site: www.wind-energie.de/en/

The German Wind Energy Association (BWE) is the largest renewable energy association in the world, with almost 19,000 members. The members consist of wind technology manufacturers, operators and their shareholders, planning office financiers, scientists, engineers, technicians, and lawyers. Members also include early conservationists, schoolchildren, and students. BWE maintains its regional roots, with 13 state and 41 regional associations providing consultation and the provision of information. Advocacy activities target political parties and decision makers at the regional and state level and provide expertise in designing policies to promote the optimum conditions for wind energy development. BWE works to implement projects with high levels of support and acceptance by all of those affected and involved. BWE is responsible for organizing regional presentations for congresses and trade fairs. It focuses on confronting criticism by local stakeholders through the provision of press releases and information presentations. BWE publishes a free monthly magazine called *New Energy,* which highlights the vast array of renewable energy sources available.

Global Wind Energy Council (GWEC)
Renewable Energy House
63-65 rue d'Arlon
B-1040 Brussels, Belgium
32 2 400 1029
Fax: 32 2 546 1944
E-mail: info(at)gwec.net
Web site: www.gwec.net/

The Global Wind Energy Council (GWEC) was established in early 2005 to provide a coordinated, representative forum for the entire wind energy sector at an international level. GWEC's mission is to promote wind power as one of the world's leading energy sources. To promote the development of wind power, the GWEC works on the adoption of policy initiatives, global outreach, education, and the dissemination of information. Members of GWEC include over 1,500 companies, organizations, and institutions in more than 50 countries and account for over 99 percent of the installed wind capacity in the world.

Integrated Waste Services Association (IWSA)
1331 H Street NW, Suite 801
Washington, DC 20005
(202) 467-6240
Fax: (202) 467-6225
E-mail: tmichaels@wte.org
Web site: www.wte.org
Contact: Ted Michaels, president

The Integrated Waste Services Association (IWSA) was founded in 1991 to provide innovative solutions to waste management problems through the use of waste-to-energy technology. Its membership base includes companies and communities that own waste-to-energy plants. Like other trade associations, the Integrated Waste Services Association engages in activities like research, political advocacy, education, and information compilation to promote the development and use of waste-to-energy technology.

International Geothermal Association (IGA)
IGA Secretariat c/o Samorka Sudurlandsbraut
48 108 Reykjavik, Iceland
354 588 4437
Fax: 354 588 4431
E-mail: iga@samorka.is
Web site: iga.igg.cnr.it/index.php
Contact: IGA Secretariat

The International Geothermal Association (IGA), founded in 1988, serves as a scientific, educational, and cultural nonprofit, nongovernmental organization, operating worldwide. There are more than 2,000 members in 65 countries. The IGA is nonpolitical and acts as a consultant to the Economic and Social Council of the United Nations, and serves as a partner of the European Union for the Campaign for Take Off (CTO) the Renewable Energy. With an objective to encourage research, development, and utilization of geothermal resources worldwide, the IGA compiles, publishes, and disseminates scientific and technical information on geothermal resources. The organization works to facilitate information exchange between geothermal experts and the general public.

Renewable Fuels Association (RFA)
One Massachusetts Avenue NW, Suite 820
Washington, DC 20001
(202) 289-3835
E-mail: info@ethanolrfa.org
Web site: www.ethanolrfa.org/

The Renewable Fuels Association (RFA) is a trade association that has represented the U.S. ethanol industry since 1981. Its membership base consists of businesses, organizations, and individuals that want to see increased ethanol use in the United States. To increase the production and use of ethanol, the group promotes state and federal legislation that will result in favorable policies, regulations, and research and development for ethanol. Additionally, the Association attempts to provide timely information for decision makers in the ethanol industry.

Solar Electric Power Association (SEPA)
1341 Connecticut Avenue NW, Suite 3.2
Washington, DC 20036
(202) 857-0898
Fax: (480) 393-5631
E-mail: info@solarelectricpower.org
Web site: www.solarelectricpower.org/

The Solar Electric Power Association (SEPA) is a nonprofit organization with members including 175 utilities, electric service providers, manufacturers, installers, government entities, and research institutions. The mission of SEPA is to encourage and facilitate the development and use of solar electric power by electric utilities and service providers and their customers. SEPA fosters business-to-business networking to encourage the sharing of information on solar electric technologies, applications, and programs. It reports to members on policies, regulations, and legislation. In addition, it provides an online resource library with over 200 solar documents that are categorized by individual topic and date. These documents include reports from around the industry that are added as they become available. SEPA also conducts studies of utility solar cases to determine the types of utility programs that facilitate solar power development and use. Once a year, it hosts the Solar Power Conference and Expo, which provides educational workshops in areas of policy, technology, markets, and finance. Additionally, it publishes a biweekly electronic newsletter that provides up-to-date news and information on solar technologies, programs, and policies.

Solar Energy Industries Association (SEIA)
805 15th Street NW, #510
Washington, DC 20005
(202) 682-0556
Fax: (202) 682-0559
E-mail: info@seia.org
Web site: www.seia.org/

Solar Energy Industries Association (SEIA) is a trade association that represents businesses and individuals from all aspects of the solar industry. Its purpose is to expand the use of solar technology across the world. For its members, it provides information pertinent to both political advocacy of solar technology and the general developments in the field, and it provides networking opportunities through conferences. The Association provides press releases for the media to keep the public informed about solar energy developments and benefits. It produces publications to inform its members about tax laws, regulations, and technologies.

U.S. Government Agencies

California Energy Commission
1516 Ninth Street, MS-29
Sacramento, CA 95814-5512
(916) 654-4287
E-mail: renewable@energy.state.ca.us
Web site: www.energy.ca.gov

The California Energy Commission is the state's primary energy policy and planning agency. Created by the legislature in 1974 and located in Sacramento, the Commission has five major responsibilities: forecasting future energy needs and keeping historical energy data; licensing thermal power plants 50 megawatts or larger; promoting energy efficiency through appliance and building standards; developing energy technologies and supporting renewable energy; and planning for and directing state response to energy emergencies. With the signing of the Electric Industry Deregulation Law in 1998 (Assembly Bill 1890), the Commission's role includes overseeing funding programs that support public interest energy research; advancing energy science and technology through research, development, and demonstration; and providing market support to existing, new, and emerging renewable technologies.

Energy Information Administration (EIA)
1000 Independence Avenue SW
Washington, DC 20585
(202) 586-8800
E-mail: InfoCtr@eia.doe.gov
Web site: www.eia.doe.gov

The Energy Information Administration (EIA), created by Congress in 1977, is a statistical agency of the U.S. Department of Energy. Its mission is to provide policy-independent data, forecasts, and analyses to promote sound policy making, efficient markets, and public understanding regarding energy and its interaction with the economy and the environment. By law, EIA's products are prepared independently of administration policy considerations. EIA neither formulates nor advocates any policy conclusions.

EIA issues a wide range of weekly, monthly, and annual reports on energy production, stocks, demand, imports, exports, and prices, and it prepares analyses and special reports on topics of current interest. Examples include weekly reports, monthly reports, annual reports, and special reports. EIA's data and analyses are widely used by researchers and policy makers from countries around the world.

Office of Energy Efficiency and Renewable Energy (EERE)
Mail Stop EE-1
Department of Energy
Washington, DC 20585
(877) 337-3463
Web site: www1.eere.energy.gov

The Office of Energy Efficiency and Renewable Energy (EERE) falls under the Department of Energy and is the administrative body that coordinates all policy efforts toward the advancement of energy efficiency and renewable energy. The EERE mission is to strengthen America's energy security, environmental quality, and economic vitality in public–private partnerships that enhance energy efficiency and productivity; bring clean, reliable, and affordable energy technologies to the marketplace; and make a difference in the everyday lives of Americans by enhancing their energy choices and their quality of life. The Office of Energy Efficiency and Renewable Energy will play an important role in the Advanced Energy Initiative, passed by President George W. Bush, by advancing technologies such as biomass and biofuels, solar power, wind power, advanced vehicles, and hydrogen fuel cells. EERE sponsors various initiatives to build awareness about energy efficiency and renewable energy topics and to coordinate efforts toward specific goals.

U.S. Department of Energy (DOE)
1000 Independence Avenue SW
Washington, DC 20585
(202) 586-5000

Fax: (202) 586-4403
E-mail: The.Secretary@hq.doe.gov
Web site: www.energy.gov
Contact: Secretary of Energy

The history of the Department of Energy (DOE) can be traced back to the Manhattan Project, when, in 1946, Congress created the Atomic Energy Commission (AEC) to control the various aspects of the project. In 1977, the Department of Energy Organization Act distinguished the DOE from the AEC and brought together for the first time all programs related to energy and to the design, construction, and testing of nuclear weapons. Over the years, the DOE has focused less on national security directly and more on energy security. Today, DOE stands at the forefront of helping the nation meet its energy, scientific, environmental, and national security goals. These include developing and deploying new energy technologies, reducing dependence on foreign energy sources, protecting the nuclear weapons stockpile, and ensuring that the United States remains competitive in the global marketplace. Recent initiatives to spur scientific innovation and technology development have served to expand DOE's continuing support for the competitive energy markets, both domestically and internationally, and for policies that facilitate continued private investment in the energy sector. In addition, DOE supports the demonstration and deployment of energy technologies through collaborative efforts with the private sector and public sector entities.

The Department of Energy's overarching mission is to advance the national, economic, and energy security of the United States; to promote scientific and technological innovation in support of that mission; and to ensure the environmental cleanup of the national nuclear weapons complex. The DOE's strategic goals to achieve the mission are designed to deliver results along five strategic themes: energy security, nuclear security, scientific discovery and innovation, environmental responsibility, and management excellence.

International Agencies

A Global Overview of Renewable Energy Sources (AGORES)
Lior International SA
P.O. Box 30-1560

Hoeilaart, Belgium
32 2 657 5300
Fax 32 2 657 3640
E-mail: agores@lior-int.com
Web site: www.agores.org

A Global Overview of Renewable Energy Sources (AGORES) can be compared to the Energy Information Administration in the United States. It is an extensive global international information center and knowledge gateway for renewable energy. The site is designed to promote the European Union's strategy to achieve 12 percent of renewable energy resources (RES) by 2010 and to reach a reduction of 20 percent of fossil fuel energy consumption by 2020. It does this by allowing fast and efficient access to an extensive range of information and connecting all the major renewable energy players. The aim is not to duplicate any information that already exists on the Internet, but simply to make efficient links to this data. The site is divided into four main sections: Policy—EC and national strategies for the increased implementation of renewable energy; Fields—information on all areas of renewable energy activity; Sectors—information on the various sources of renewable energy; and an extensive who's who directory of key players from around the world. AGORES also contains AL-TENER (alternative energy) Projects, a searchable renewable and alternative energy project database and an extensive list of recommended Web sites.

European Renewable Energy Council (EREC)
Renewable Energy House
63-67 rue d'Arlon
B-1040 Brussels, Belgium
32 2 546 1933
Fax: 32 2 546 1934
Web site: www.erec.org
E-mail: erec@erec.org

The European Renewable Energy Council (EREC) was created in 2000. EREC serves as the umbrella organization for the European renewable energy industry, trade, and research associations. Members are active in the sectors of bioenergy, geothermal, ocean, small-scale hydropower, solar electricity, solar thermal, and wind energy. The following nonprofit organizations and federations are members: European Biomass Association, European Geothermal

Energy Council, European Photovoltaic Industry Association, European Small Hydropower Association, European Solar Thermal Industry Federation, European Biomass Industry Association, European Wind Energy Association, European Renewable Energy Centres Agency, European Bioethanol Fuel Association, European Renewable Energies Federation, and European Ocean Energy Association. EREC acts as a forum for information exchange and discussions on issues related to renewable energy. It provides information and consulting services on renewable energy to political leaders at the local, regional, national, and international levels in order to promote the adoption of policies that create favorable conditions for renewable energy development. To reach these objectives, EREC works on a series of international projects and organizes regular conferences, workshops, and events. In addition, EREC produces a series of position papers and publications on renewable energy–related topics.

International Energy Agency (IEA)
9 rue de la Fédération
75739 Paris Cedex 15, France
33 1 40 57 65 00/01
Fax: 33 1 40 57 65 59
E-mail: info@iea.org
Web site: www.iea.org

Based in France, the International Energy Agency (IEA) acts as energy policy adviser to 26 member countries in an effort to ensure reliable, affordable, and clean energy for their citizens. Founded during the oil crisis of 1973–1974, the IEA's initial role was to coordinate measures in times of oil supply emergencies. As energy markets have changed, so has the IEA. Its mandate has broadened to incorporate the so-called Three Es of balanced energy policy making: energy security, economic development, and environmental protection. Current work focuses on climate change policies, market reform, energy technology collaboration, and outreach to the rest of the world, especially major producers and consumers of energy like China, India, Russia, and the OPEC countries. With a staff of around 150, mainly energy experts and statisticians from its 26 member countries, the IEA conducts a broad program of energy research, data compilation, publications and public dissemination of the latest energy policy analysis, and recommendations on good practices.

International Solar Energy Society (ISES)
Villa Tannheim Wiesentalstr. 50
79115 Freiburg, Germany
49 761 45906 0
Fax: 49 761 45906 99
E-mail: hq@ises.org
Web site: www.ises.org/

The International Solar Energy Society (ISES) was founded in 1954 to serve the needs of the solar energy community. It is a UN-accredited nongovernmental organization (NGO) operating in more than 50 countries. The organization supports its members by helping to advance the development and use of renewable energy technology. To achieve this, ISES educates about the benefits of renewable energy all over the world. The organization emphasizes the importance of a global community and social responsibility in its advocacy of sustainable energy choices. To do this, it works to forge alliances among industries, individuals, and institutions in support of renewable energy technologies, through communication, cooperation, support, and exchange. It supports the development of renewable energy through applying practical projects, encouraging technology transfer, conducting education and training, and supporting global energy development. It also supports the science of solar energy through the encouragement of research in the field with an eye toward community growth. The organization also provides access to information through the use of the modern technology in order to keep its members updated and connected.

Renewable Energy and Energy Efficiency Partnership (REEEP)
Vienna International Centre, Room D1732
Wagramerstrasse 5
A-1400 Vienna, Austria
43 1 26026 3425
Fax: 43 1 21346 3425
E-mail: info@reeep.org
Web site: www.reeep.org

The Renewable Energy and Energy Efficiency Partnership (REEEP) is an active, global, public–private partnership that structures policy and regulatory initiatives for clean energy and that facilitates financing for energy projects. Backed by more than 200 national governments, businesses, development banks, and

NGOs, REEEP contributes to international, national, and regional policy dialogues. The aim behind the organization is to accelerate the integration of renewables into the energy mix and to advocate energy efficiency as a path to improved energy security and reduced carbon emissions, ensuring socioeconomic benefits. With a network of eight regional secretariats and more than 3,500 members, REEEP has the ability to effect change worldwide. The partnership has funded more than 50 high-quality projects in 44 countries that address market barriers to clean energy in the developing world and economies in transition. These projects are beginning to deliver new business models, policy recommendations, risk mitigation instruments, handbooks, and databases.

Renewable Energy Policy Network for the 21st Century (REN21)
15 rue de Milan
75441 Paris Cedex 9 France
33 1 44 37 50 94
Fax: 33 1 44 37 50 95
E-mail info@ren21.net
Web site: www.ren21.net
Contact: REN21 Secretariat

Renewable Energy Policy Network for the 21st Century (REN21) is a global policy network in which ideas are shared and action is encouraged to promote renewable energy. It provides a forum for leadership and exchange in international policy processes. It bolsters appropriate policies that increase the wise use of renewable energies in developing and industrialized economies. Open to a wide variety of dedicated stakeholders, REN21 connects governments, international institutions, nongovernmental organizations, industry associations, and other partnerships and initiatives. Linking actors from the energy, development, and environment communities, REN21 leverages their successes and strengthens their influence for the rapid expansion of renewable energy worldwide.

World Renewable Energy Congress/Network (WREC/WREN)
P.O. Box 362
Brighton BN2 1YH, United Kingdom
44 0 1273 625643
Fax: 44 0 1273 625768

E-mail: asayigh@netcomuk.co.uk
Web site: www.wrenuk.co.uk/
Contact: Prof. A. Sayigh

World Renewable Energy Congress/Network (WREN) is a major nonprofit organization registered in the United Kingdom with charitable status and affiliated to the United Nations Educational, Scientific, and Cultural Organisation (UNESCO), the deputy director general of which is its honorary president. It has a governing council, an executive committee, and a director general. It maintains links with many United Nations, governmental, and nongovernmental organizations. Established in 1992 during the second World Renewable Energy Congress in Reading, United Kingdom, WREN is one of the most effective organizations in supporting and enhancing the utilization and implementation of renewable energy sources that are both environmentally safe and economically sustainable. This is done through a worldwide network of agencies, laboratories, institutions, companies, and individuals, all working together toward the international diffusion of renewable energy technologies and applications. Representing most countries in the world, it aims to promote the communication among and technical education of scientists, engineers, technicians, and managers in this field and to address itself to the energy needs of both developing and developed countries. Over $2 billion have been allocated to projects dealing with renewable energy and the environment by the World Solar Summit and World Solar Decade along with the World Bank.

8

Resources

This chapter provides a list of resources for anyone who wants to delve deeper into the topic of alternative and renewable energy. The first section contains print resources, including books, articles, and journals. The next section provides the nonprint resources that are widely used for the purposes of researching energy. The nonprint resources are divided into subsections on DVDs, Internet resources, and databases and CD-ROMs. Some sources provide general alternative and renewable energy information, and others are specific to certain types of energy. Although many sources are neutral and scientifically oriented, some make a case for the use of alternative and renewable energy. Thinking critically about the information provided is essential to coming to independent conclusions about the issues. Many of the sources included in this chapter are accessible to people without strong math and science backgrounds, but they can still offer important information for those with technical experience.

Print Resources

Books and Articles

Asmus, Peter. *Reaping the Wind: How Mechanical Wizards, Visionaries and Profiteers Helped Shape Our Energy Future.* Washington, D.C.: Island Press, 2001

Asmus follows the development of wind technology from a nascent energy source to its current unprecedented growth in the mainstream energy market. He provides a historical account of the growth of wind power and of the inventors, scientists, and entrepreneurs who helped it achieve its current rates of use. Information about the technologies and policies of wind power is also included with this interesting and informative human interest story.

Berman, Daniel M. *Who Owns the Sun? People, Politics, and the Struggle for a Solar Economy.* **White River Junction, VT: Chelsea Green Publishing, 1996**

Berman provides an overview of the reasons that solar energy should be included in the current energy mix and why it is not. He exposes the power struggle between renewable energy advocates and the fossil fuel industry to explain the low rate of solar energy development. The book provides suggestions for overcoming the problems contributing to this lack of development, including the public ownership of utilities and the removal of government subsidies for fossil fuels.

Congressional Research Service. *Energy from the Ocean.* **Honolulu, HI: University of the Pacific Press, 2002**

This book is a comprehensive overview of the different types of ocean energy sources. The chapters cover wave energy, ocean thermal energy conversion, tidal energy, energy from currents, and energy from oceanic winds. Other nonrenewable ocean energies are also addressed. The information is accessible to those who lack expertise in the energy fields.

Geller, Howard. *Energy Revolution: Policies for a Sustainable Future.* **Washington, D.C.: Island Press, 2002**

Geller argues that current government policies are the main hindrance to a transition away from fossil fuels to renewable energy sources. This book examines policy options for reducing our dependence on fossil fuels and helping to achieve more sustainable energy solutions, including energy efficiency and renewable energy use. Geller provides an overview of the current situation, a look at the barriers that need to be overcome, 10 case studies on policies that have worked in other parts of the world, suggestions for the improvement of international policies and institutions, and evidence that a shift to a more sustainable energy economy is possible. This book is useful for anyone seeking a critical assessment of current energy policies.

Goettemoeller, Jeffrey, and Adrian Goettemoeller. *Sustainable Ethanol: Biofuels, Biorefineries, Cellulosic Biomass, Flex-fuel Vehicles and Sustainable Farming for Energy Independence.* Maryville, MO: Prairie Oak Publishing, 2007

Goettemoeller and Goettemoeller give an in-depth account of the pros and cons of ethanol use in North America. Providing current information on the state of ethanol, the authors show that ethanol technology is becoming more efficient and less dependent on fossil fuel inputs. They focus on the advancements in ethanol production that have been made with the intention of overcoming the criticisms of using ethanol as a fuel, including the emergence of sustainable farming (so that ethanol crops do not compete with our food supply) and the development of cellulosic ethanol (produced from agricultural waste). Anyone interested in learning more about this complex energy source will appreciate this useful and easy-to-understand book.

Gupta, Harsh K., and Sukanta Roy. *Geothermal Energy: An Alternative Resource for the 21st Century.* Burlington, MA: Elsevier Science, 2006

Gupta and Roy present a comprehensive overview of geothermal energy from a multidisciplinary perspective. They address geology, geophysics, engineering, and social perspectives of geothermal development, resting on the premise that the world needs to fully utilize this energy source. The book's information on geothermal sources is up-to-date. This is an excellent resource for anyone wanting to learn more about geothermal energy.

Hoffman, Peter. *Tomorrow's Energy: Hydrogen, Fuel Cells, and the Prospects for a Cleaner Planet.* Cambridge, MA: MIT Press, 2001

Hoffman provides a detailed overview of the concept of a hydrogen economy. Arguing in favor of the need to move beyond fossil fuels to sustainable energy sources, Hoffman presents the historical, technological, and economic considerations related to hydrogen as an energy carrier. He provides information on the chemical characteristics of hydrogen and the history of its development. He looks at a variety of contexts in which hydrogen could be used, including transportation, electric generation, and utility applications. He discusses fuel cells and the technological and political problems that are hindering their progress, stressing the need to include them in the hydrogen economy. This book provides an in-depth yet easy-to-understand overview of hydrogen for anyone interested in the subject.

Komor, Paul. *Renewable Energy Policy.* **Lincoln, NE: iUniverse, 2004**

Komor analyzes government efforts to promote renewable energy in several different countries, taking lessons from each. He addresses the technological, economical, and political aspects that contribute to the success or failure of renewable energy policies. This book is clear and simple enough for those without expertise in policy to understand. It gives an excellent overview of general types of policies that can be used for promoting renewable energy, while providing specific information in the case studies on Germany, the United Kingdom, and the United States.

Kryza, Frank T. *The Power of Light: The Epic Story of Man's Quest to Harness the Sun.* **Columbus, OH: McGraw-Hill, 2003**

Kryza examines both the social and the technological history of solar power. He follows the development of solar technology by chronologically detailing the inventions that contributed to our ability to harness the sun's power. He provides a detailed account of the inventors and scientists who dedicated their lives to the advancement of solar energy technology. The overarching theme of the book is to convey the revolutionary impact of scientific advancements.

Leggett, Jeremy. *The Empty Tank: Oil, Gas, Hot Air, and the Coming Global Financial Catastrophe.* **New York: Random House, 2005**

Leggett combines several pertinent topics related to alternative and renewable energy and develops a well-supported argument for less reliance on fossil fuels. He provides convincing evidence that oil reserves should not be taken for granted. He reviews the major arguments about the seriousness of global warming and the real possibility of economic collapse due to its effects. Despite the severity of the doomsday scenarios presented, Leggett provides hope with his list of solutions. He argues for the use of wind and solar, and he suggests that, if production efforts like those made during World War II were used for renewable energy, we would have no problem overcoming the challenges we face. This book is a good source for people who are interested in the arguments against fossil fuel use and for those with an interest in global warming.

Scheider, Walter. *A Serious but Not Ponderous Book on Nuclear Energy.* **Manchester, UK: Cavendish Press, 2001**

This is an accessible explanation of renewable energy for anyone with an understanding of high school mathematics and algebra. Scheider provides a clear, engaging, and unbiased scientific look at nuclear energy. His goal is to make nuclear energy principles understandable so that people can make their own judgments on the viability of nuclear energy. This book covers the sources of nuclear energy, the different nuclear energy reactor designs, the inner workings of a nuclear reaction, and the nature of radioactivity and its effects. Mainly a technical explanation of nuclear principles, the book does not provide much information on the social and political issues surrounding nuclear power.

Solomon, Barry D., and Abhijit Banerjee. "A Global Survey of Hydrogen Energy Research, Development and Policy." *Energy Policy* **34, 7 (May 2006): 781–792.**

Solomon and Banerjee raise "serious questions about the sustainability of a hydrogen economy" through a survey of the current measures being taken to implement hydrogen energy use in developing and developed nations around the world. Not only do they look at the status of research and development projects and public policies regarding hydrogen, they address the energy mix needed in these countries to initiate a hydrogen economy. They find in their survey that only two major automakers and two nations have time tables for the production of hydrogen fuel cells or vehicles. Additionally, their survey determines that only two countries, Brazil and Iceland, would be able to produce the energy needed for a hydrogen economy with renewable energy sources.

Wizelius, Tom. *Developing Wind Power Projects: Theory and Practice.* **London: Earthscan Publications, 2007**

This book is geared toward students and professionals rather than engineers or policy analysts. The purpose is to provide an overview of how wind power projects interact with society economically, socially, and politically. The book also provides a comprehensive overview of wind power basics to provide the foundational information necessary for understanding how it works. Mainly, the book views wind power in the realistic context of project development.

Journals and Magazines

Environmental Science and Technology

This journal provides scientific and technical information on a wide range of environmental disciplines. It often publishes studies on specific renewable energy technologies, ranging from their technical viability to their social and economic consequences. The journal is published by the American Chemical Society.

Nuclear Technology

This journal is published by the American Nuclear Society. *Nuclear Technology* is an international publication that reports new information in all areas of the practical application of nuclear science. Topics include all aspects of reactor technology: operations, safety materials, instrumentation, fuel, and waste management. Also covered are medical uses, radiation detection, production of radiation, health physics, and computer applications.

Renewable Energy

This international journal, published by the World Renewable Energy Network, seeks to promote and share knowledge of the various topics and technologies of renewable energy. The target audience includes researchers, economists, manufacturers, and world agencies. The goal is to help the target audience and societies stay informed about the latest developments in the various renewable energy fields as well as to unite in finding alternative energy solutions to current issues such as the greenhouse effect and the depletion of the ozone layer. The scope of the journal encompasses photovoltaic technology conversion, solar thermal applications, biomass conversion, wind energy technology, materials science technology, solar and low energy architecture, energy conservation in buildings, climatology and meteorology (geothermal, wave and tide, ocean thermal energies, mini hydro power and hydrogen production technology), socioeconomic issues, and energy management.

Renewable Energy World

This online magazine is published bimonthly for renewable energy professionals. The international publication contains the latest news and features on all aspects of renewable energy, including in-depth policy and market updates, project profiles, technology reports, company listings, information on renewable energy conventions and exhibitions, and more. Much of the information is useful for nonprofessionals as well. Most articles cover the mainstream renewable energies like wind, solar, biomass, and biofuels.

Renewable and Sustainable Energy Reviews

This journal publishes specially commissioned review articles designed to bring together, under one cover, current advances in the ever broadening field of renewable and sustainable energy. The coverage of the journal includes all of the renewable energy resources discussed in Chapter 1; information on the applications of renewable energy in the building, industry, electricity, and transportation sectors; policy aspects of renewable energy; environmental impact and sustainability assessments; and regionally focused coverage. The journal is targeted at scientists, researchers, and consultants involved in all aspects of renewable energy.

Nonprint Resources

DVDs

Energy Crossroads: A Burning Need to Change Course
Type: DVD
Length: 54 Minutes
Date: 2007
Cost: $25.95
Source: Tiroir A Films

Energy Crossroads highlights the possibility of an energy shortage due to increases in global energy demand and resource depletion. This film provides viewpoints from experts in an easy-to-understand and convincing manner. The conclusion is that, to avert an energy crisis, renewable energy sources must be developed as soon as possible.

Inconvenient Truth, An
Type: DVD
Length: 94 minutes
Date: 2006
Cost: $19.99
Source: Paramount Classics

An Inconvenient Truth is an award-winning documentary about Al Gore's efforts to educate the public about global warming and options for dealing with it. In the movie, Gore gives a slide show on the scientific evidence supporting the claim that global warming is being caused by carbon dioxide emissions from human activities and the effects of climate change. The politics and economics of global warming are discussed, as well as the refutations of critics. Renewable energy is discussed as a solution for emitting less carbon dioxide into the atmosphere. This DVD makes global warming arguments easy to understand.

NOVA: Solar Energy—Saved by the Sun
Type: DVD
Length: 56 minutes
Date: 2007
Cost: $17.99
Source: WBGH Boston

NOVA presents the latest thinking by solar advocates and skeptics as it investigates cutting-edge research developments. The film introduces viewers to the scientists and businesspeople who are working to make solar power practical—for lighting, heating, and running power plants. This movie is based on the premise that there is an impending energy crisis and that it is time to start seriously pursuing solar energy as an option for meeting our power needs. The program provides an accessible look at solar technology and its prospects for the future.

Power Shift: Energy + Sustainability
Type: DVD
Length: 26 minutes
Date: 2004

Cost: $24.95
Source: Green Planet Films

Hosted and narrated by Cameron Diaz, this four-part program takes place around the world, exploring all the ways that energy is consumed. It examines energy use in the lives of astronauts in the International Space Station, villagers in the Amazon River Basin, and an actress in Hollywood. In the first part, "Connections," linkages between our energy consumption and the global community are explored. The second part, "Cradle to Cradle," chronicles William McDonough's innovative green building designs. The third part, "Energy Path," follows energy from its end use to its source. The final part, "Be the Difference," encourages personal responsibility and action for reducing energy consumption and seeking out sustainable energy options. The movie is geared toward students, suggesting how they can create a sustainable future while examining critical energy issues.

Internet Resources

Alternative Energy News
www.alternative-energy-news.info/

The Alternative Energy News Web site provides links to news and information about renewable energy technologies. News is categorized by energy type, and there is also a link to energy news in general. News from multimedia sources gives visitors a choice of reading articles or watching news clips. Subjects are subdivided even further by categories ranging from batteries to ethanol to politics, making the site a very comprehensive source of information. A directory of over 1,500 sites on renewable energy is also available. Discussion forums allow visitors to share ideas and opinions about renewable energy issues. This page is great for students and other nontechnical readers interested in broadening their knowledge of renewable energy issues.

Canadian Renewable Energy Network (CanREN)
www.canren.gc.ca/default_en.asp

The Canadian Renewable Energy Network (CanREN) offers general information on renewable energy sources and highlights the technologies and applications being developed to harness these sources. It seeks to increase the understanding of renewable energy so that development and commercialization of renewable energies will be accelerated. The Web site allows visitors to browse information by energy source and other topics. There are links to other renewable energy Web sites, educational and instructional materials, a glossary, maps, and other data. Although mostly based on Canadian efforts, the Web site provides a wealth of general information that visitors from any country will find useful.

Energy Information Administration (EIA)
www.eia.doe.gov/

The EIA Web page provides energy statistics on traditional and renewable energy use in the United States and around the world. Statistics on the environment are also available. The renewables covered include biomass, solar, wind, geothermal, hydro, and ethanol. Data is divided into categories including U.S. data, international data, reports, analyses, and forecasts. Summaries on historical data are available. The site includes numerous color graphs and charts, making the information interesting and easy to understand. Reference materials and a glossary are provided for clarification.

European Commission, Energy
ec.europa.eu/energy/index_en.html

The Energy Web page is maintained by the European Commission for the European Union (EU) and focuses on energy and transportation. The site contains information on the energy policies of the EU and its member states, transportation issues and statistics, and reports related to both. The *Annual Energy and Transport Review* can be accessed from this page. Renewable energy is portrayed on this site as a solution to energy and transportation issues in the EU.

Geothermal Resources Council (GRC)
www.geothermal.org

The Geothermal Resources Council Web page provides access to several resources for information on geothermal energy, including its extensive library. The GRC's library is the largest geothermal technical library in the world, offering title and author citations for over 30,000 papers, articles, books, and other geothermal publications. The library is updated regularly and can be navigated using keyword searches. There is also access to information about geothermal power production in the United States. This is an excellent site for in-depth information on geothermal technology.

A Global Overview of Renewable Energy Sources (AGORES)
www.agores.org/

The AGORES site is an extensive global international information center and knowledge base for renewable energies. Although it covers information on the entire world, its main focus is on European development and experience in the renewable energy field. It is updated daily and is divided into four main sections: policy, fields, sectors, and who's who. The policy section covers developments in national and international renewable energy policy. The fields section provides information on all areas of renewable energy activity. The sectors section provides information on the various sources of renewable energy. Who's who covers key players and stakeholders from around the world, such as government agencies and associations. AGORES also contains several databases on projects, Web site links, publications, and news about renewable energy.

Green Energy News
www.nrglink.com/

Green Energy News is a Web site that offers news and commentary about renewable energy as it relates to business, technology, issues, and policy. It is geared toward a wide audience, including consumers, industry affiliates, educators, and policy makers. Topics include green energy building design, renewable fuels, environmental issues, social issues, political issues, electricity generation, and transportation. The Web site also provides a link to funding opportunities for renewable energy projects.

International Energy Agency (IEA)
www.iea.org/

The IEA Web page is devoted to international energy information. The information links are conveniently divided by topic, country, and publications and papers. Statistical information on most countries in the world is available, much of it free of charge and available to the public. The *World Energy Outlook,* the annual report on global energy issues published by the IEA, is available on this Web page. Maps and graphs are used to make the information more accessible. Information on the environment is also available. The Web site offers several links to additional information on numerous topics and to official government Web pages.

National Renewable Energy Laboratory (NREL)
www.nrel.gov

The NREL Web page primarily provides information on research projects that the laboratory is conducting, with great emphasis on the technological side of renewable energy. The site also contains general information about wind, solar, biomass, and geothermal energies for consumers, homeowners, businesses, and students, with a special emphasis on advanced renewable energy fuels. Pertinent news information about the laboratory, general news about renewable energy, and renewable energy resource maps are also available. Links to the National Wind Technology Center and the Solar America Initiative are found at this page.

Renewable Energy Atlas of the West
www.energyatlas.org

The *Renewable Energy Atlas of the West* is an 80-page, full-color atlas of renewable energy sources in the Western United States. The atlas is a synthesis of several energy resource maps for solar, wind, geothermal, and biomass. It not only provides quantitative data on estimates of untapped renewable energy resources, but it highlights the issues surrounding their development. It serves a variety of purposes for several stakeholders, including the facilitation of planning and understanding of areas for where potential is unmet. It is freely accessible to the public and provides key data on renewable energy potential.

University of California Energy Institute (UCEI)
www.ucei.berkeley.edu/

The UCEI Web page provides access to data that the institute has compiled and publications of its researchers. It is a good source for information on energy markets. Although not much general information is provided, the site has a comprehensive list of links to other energy research organizations ranging from universities to governmental departments to the private sector. It is a good starting place for information on research efforts that may not have been published yet.

U.S. Department of Agriculture (USDA), Biofuels Information
www.nal.usda.gov/ttic/biofuels.htm

The USDA Web page on biofuels information provides general information on biofuels as well as news on biofuel events, legislation, and policy reports. Funding opportunities for research and commercial endeavors are also listed on the Web site. Access to research on results for biofuel studies is available, and links to other helpful Web sites on biofuels are provided.

U.S. Department of Energy (DOE) Hydrogen Program
www.hydrogen.energy.gov/

The DOE Hydrogen Program Web page is an excellent resource for information on hydrogen. The Program's efforts are discussed, but the page is mostly devoted to general information on hydrogen production, delivery, storage, and fuel cells. There is a section on international collaborative efforts to research hydrogen. Links to other government programs that are also researching hydrogen are provided in the general information sections. This is a good starting place for research on government efforts to develop hydrogen.

U.S. Department of Energy, Office of Energy Efficiency and Renewable Energy (EERE)
www.eere.energy.gov/

The Web page for the DOE's Office of Energy Efficiency and Renewable Energy contains general information about renewable energies in addition to information about specific programs to develop renewable energy in the United States. Information is divided by energy source and provides links to inform stakeholders, policy makers, consumers, students, and researchers. Information is available on renewable energies that are not yet widely used, like wave and tidal energy, in addition to established renewable energy technologies. There are also links to data on specific energy sources and to published resources. Glossaries and databases are provided to improve understanding of the complexities of renewable energy resources. The EERE's green power network can be accessed from this page. The green power network provides information on the availability of renewable energy in different states for consumers. The alternative fuels and advanced vehicles data center program, which provides information to help reduce oil dependency through the use of alternative fuels and advanced vehicle technologies, can also be accessed on this page.

U.S. Environmental Protection Agency Landfill Methane Outreach Program (LMOP)
www.epa.gov/lmop/

The LMOP home page is a good source of information for landfill methane capture. The site contains information on the program and general information on landfill methane energy sources. Colorful diagrams illustrate how methane is captured. There are also photographs of the different types of wells used to harness methane. Maps of methane capture projects are provided. Statistics on energy production and benefits of methane capture are addressed.

Virtual Nuclear Tourist
www.nucleartourist.com/

This Web site was developed by Joseph Gonyeau, an independent energy consultant, who has worked at several nuclear energy plants. Although the author is biased in favor of nuclear energy, the site provides a broad range of information. Hot topics are addressed, photographs from inside several nuclear plants are available, maps detailing the location of nuclear power plants in the United States and around the world can be accessed, comparisons of nuclear power with other energy sources are made, the pros and cons of nuclear power are discussed, and an overview with sections on how nuclear energy works and the designs of different plants is available. This site also contains information about important historical nuclear events and the Nobel laureates who contributed to the development of nuclear power. This site is a great starting place for those who want to broaden their understanding of nuclear energy.

Databases and CD-Roms

21st Century Complete Guide to Solar Energy and Photovoltaics (2004)
www.connectedglobe.com/cgi-local/amazon/cgapf.cgi?Operation=ItemLookup&ItemId=1422014681&templates=millennium

This CD-Rom from *World Spaceflight News* provides over 20,000 pages of official federal documents on the subject of solar energy and photovoltaics. It draws from government sources such as the United States Department of Energy Photovoltaics Program; National Renewable Energy Laboratory (NREL) Center for Photovoltaics; Photovoltaics for Buildings; Photovoltaics Research and Development; Photovoltaics Silicon Materials Research; Solar Decathlon; Amorphous Silicon; and NREL Solid State Theory Group and Solid State Spectroscopy. This CD-Rom provides detailed information on the programs to promote solar energy as well as technical information to give readers a more in-depth understanding of the science behind solar power.

Country Nuclear Power Profiles (CNPP)
www-pub.iaea.org/MTCD/publications/PDF/cnpp2004
/CNPP_Webpage/pages/index.htm

The CNPP Web site offers a database and a technical publication containing information highlighting the energy and economic conditions, the energy and the electricity sector, and the primary organizations involved in nuclear power in the International Atomic Energy Agency (IAEA) member states. The CNPP provides background information on the status and development of nuclear power programs in countries having nuclear plants in operation and/or plants under construction. It reviews the organizational and industrial aspects of nuclear power programs in participating countries, and it provides information about the relevant legislative, regulatory, and international environments in each country. The CNPP places the information of the current issues in the new environment in which the electricity and nuclear sectors operate, taking into account energy policy, privatization and deregulation, the role of government, nuclear energy and climate change, and safety and waste management, all of which often differ from country to country.

Database of State Incentives for Renewables and Efficiency (DSIRE) Database
www.dsireusa.org

This database is run by North Carolina State University for the North Carolina Solar Center and the Interstate Renewable Energy Council (IREC). It is funded by the U.S. Department of Energy and updated daily. The database is a comprehensive source of information on state, local, utility, and federal incentives that promote renewable energy and energy efficiency. The database is very user-friendly and has a glossary to explain various policy terms. There is also a library of technical papers, reports, and presentations written by DSIRE staff members since 1997.

EnergyFiles
www.osti.gov/energyfiles/

EnergyFiles contains over 500 databases and Web sites with information and resources related to the science and technology of interest to the Department of Energy, with particular emphasis on the physical sciences. The site is sponsored, developed, and maintained by the Department of Energy's Office of Scientific and Technical Information (OSTI). EnergyFiles combines information, tools, and technologies to facilitate access to and use of scientific resources. The R&D Highlights section provides a collection of non-technical summary information about selected research conducted or sponsored by the U.S. Department of Energy and its national laboratory infrastructure, making the Web site's contents accessible to those without a significant physical science background.

Energy Technology Data Exchange
www.etde.org/

The Energy Technology Data Exchange is a database that was established under the International Energy Agency to provide governments, industry, and the research community in the member countries with access to the widest range of information on energy research, science, and technology and to increase dissemination of this information to developing countries. The database is the largest collection of energy-related research and technology literature in the world. It contains bibliographic references to and abstracts of journal articles, reports, conference papers, books, Web sites, and other miscellaneous document types. The database is updated regularly and information comes from various reputable sources.

Global Energy Marketplace (GEM)
www.acca21.org.cn/cres.html

The GEM database consists of more than 2,500 annotated Web links on energy efficiency and renewable energy. It is sponsored by the U.S. Environmental Protection Agency to disseminate information on sustainable energy solutions to promote more responsible energy use. The database includes useful case studies, reports, publications, economic analyses, product directories, discussion groups, country profiles, mitigation assessments, and other beneficial resources.

Johannesburg Renewable Energy Coalition (JREC) Policies and Measures Database
ec.europa.eu/environment/jrec/index_en.htm

The JREC Policies and Measures Database provides information on renewable energy policies in various countries and on the renewable energy targets that they are seeking. The database is very user-friendly. Visitors can click on a country on a world map to bring up that country's renewable energy policies and measures. Full explanations of the policies are provided, along with links to outside resources, including, if applicable, the energy Web page of the government for that country.

Regional Energy Database
www.serd.ait.ac.th/red/RedHome.htm

The Regional Energy Database contains several online databases, mostly based on renewable energy efforts in Asia and the Pacific region. The Web site contains a comparison of renewable energy and energy efficiency policies and incentives, success factors, and barriers to success in the study countries (those that respond to the survey). These databases' subjects include government support, incentives, and policies; success stories; national databases of the pilot countries; and a directory of renewable energy and energy efficiency databases. The database on government support, incentives, and policies provides information on national energy policies that address the utilization of renewable energy. The database on success stories on renewable energy and energy efficiency provides examples of successful renewable energy and energy efficiency policies and developments in Asia and the Pacific region. The national databases provide country profiles highlighting the current status of renewable energy and energy efficiency in the country; policies to promote research, development, and technology sharing; and the key players in the promotion and utilization of these programs. The Regional Energy Database also contains an online directory of other existing databases on renewable energy and energy efficiency worldwide that focus on Asia and the Pacific region. The Regional Energy Database is sponsored by the United Nations Educational, Scientific, and Cultural Organization (UNESCO).

World Energy Outlook Database
www.iea.org/textbase/pm/?mode=weo

The International Energy Agency's World Energy Outlook database is a comprehensive source of various energy-related policies. It is based on information from a number of sources, including IEA Climate Change Mitigation Database, IEA Energy Efficiency Database, IEA Global Renewable Energy Policies and Measures Database, the European Conference of Ministers of Transport (ECMT) from "Transport CO_2 Policy Review Report," as well as contacts from industry and government. There is much information on the efforts of countries throughout the world to develop alternative and renewable energy and the likely outcomes of these efforts.

Glossary

advanced renewable energy tariffs (ARTs)   *See* electricity feed-in laws.

alternative energy   A category of energy that includes sources other than fossil fuels that are not considered renewable. Nuclear power and energy stored in hydrogen are the main alternative energy sources examined in this book.

anthropogenic   Term used to describe effects caused by human activities.

biomass   An energy source that consists of virtually any living plant material, as well as organic wastes from other sources, such as humans and plants.

central solar power (CSP)   Also known as a concentrating system. Sunlight is focused or concentrated to generate steam, which is then used to run turbogenerators. There are three types of CSP systems. (1) Parabolic trough systems use concentrators that focus sunlight into a tube that runs the length of the trough. (2) Towers collect sunlight using fields of sun-tracking mirrors known as heliostats and concentrate the heat onto a centrally located tower-mounted receiver where it heats a liquid, which is constantly circulated. (3) Parabolic dishes, or dish-engine systems, concentrate sunlight onto a focal point in the middle of a dish, where the heat is then transferred to an engine and converted to mechanical energy. All of these technologies can be used to heat and cool air or water and to generate electricity.

cold fusion   A theoretical type of fusion requiring a nuclear reaction in which the elements do not need to be heated to such extreme temperatures. Also known as low-energy nuclear reactions (LENR). *See also* nuclear fusion.

dish engine systems   *See* parabolic dishes.

economies of scale   The level of production at which the more goods that are produced, the less it costs to produce them.

electricity feed-in laws Laws that reflect a type of nonmarket-based policy and that require utilities to pay higher prices for electricity produced by renewable and alternative energy companies in order to stimulate investment in renewable energy technologies.

externality In economics, a cost or benefit that is not calculated into the costs or profits of production. Often such costs and benefits are paid for or enjoyed by the public and the government rather than by the industry. Negative externalities are associated with costs, and positive externalities are associated with benefits.

fast breeder reactor (FBR) A type of nuclear reactor that uses plutonium as well as uranium to fuel the reactions, they are capable of getting 60 times more energy from uranium than other reactors.

fossil fuels Fuels that are hydrocarbon based, that are derived from the fossilized remains of decayed plants and animals, and that have been captured in geological deposits. They are considered nonrenewable energy sources because they take millions of years to form and cannot be easily replenished after use. Coal, oil, and natural gas are all fossil fuels.

geothermal energy Heat that is stored in the Earth's crust, usually beneath the ground, and that is created by the subsurface decay of radioactive minerals. Hot springs and geysers are examples of geothermal energy that has traveled aboveground through the medium of water or steam.

green certificates Certificates that are granted for environmental attributes produced by green energy generators and that can then be traded on the secondary market. Also known as tradable renewable certificates (TRCs), green tags, and renewable obligation certificates (ROCs).

green energy market development A policy that allows consumers to choose between green and brown energy. Green energy includes all renewable and alternative energy sources, and brown energy refers to energy derived from fossil fuels. Under this policy, companies are required to give consumers a choice between these two types of energy, and then they are required to produce as much electricity with green energy as is demanded by consumers.

greenhouse gases (ghgs) Gases, like carbon dioxide and methane, that have the effect of trapping heat from the sun in the atmosphere, as in a greenhouse, and over time creating a rise in temperature that will have dramatic effects on ecosystems.

green tags *See* green certificates.

horizontal axis wind turbine (HAWT) The most common type of wind turbine, similar in shape to propellers.

Hubbert's curve An accurate prediction for the United States' oil topping point made by M. King Hubbert, a geophysicist who worked for

Shell Oil Company. In 1956, by combining the principles of geology, physics, and mathematics, he was able to accurately predict that the United States would reach its peak of oil production in the 1970s. His prediction was presented on a graph and was known as "Hubbert's curve" due to its shape.

hydrocarbon-based fuels   *See* fossil fuels.

large-scale hydro projects   Large hydropower systems, generally large dams, whose water-driven turbines produce electricity from flowing water.

light water graphite reactor (RBMK)   A type of nuclear reactor found only in Russia.

low-energy nuclear reactions (LENR)   *See* cold fusion.

micropower   *See* small-scale hydro projects.

net energy producer   An energy source that yields more energy than it uses.

net metering   An energy policy that allows electricity customers to generate electricity from renewable energy and use it to offset electricity they would have bought at retail prices from utility companies. Essentially, the energy meter rolls back when the customers generate more electricity than they are using.

NIMBY (not in my backyard)   An opposition to a project that benefits many people by those who will suffer the negative effects of having it near their residence. It can also be referred to as NIMBYism. Most people who have an opposition to a project that is in close proximity to them would not oppose it if it was in someone else's backyard.

nuclear fission   A thermonuclear process in which energy is harnessed from the splitting of atoms by placing nuclear fuels into a nuclear reactor core, where a self-sustaining process of atom splitting occurs and generates heat. The heat is then used to create steam, which then powers a turbine to generate electricity.

nuclear fusion   Energy derived from the fusing of light elements to create heavy elements. Nuclear fusion is the dominant reaction in thermonuclear weapons, but it has not been used to generate energy.

ocean thermal energy conversion (OTEC)   A type of ocean energy in which power is extracted from temperature differences between the surface of the ocean and deep waters through the use of a heat engine.

parabolic dishes   A type of central solar power system in which sunlight is concentrated onto a focal point in the middle of a dish, where the heat is then transferred to an engine and converted to mechanical energy. Also known as dish engine systems or solar dish systems.

parabolic trough A type of central solar power system that uses concentrators to focus sunlight into a tube that runs the length of the trough.

passive solar A form of solar heating that is utilized when buildings are designed in a way that captures heat and transmits daylight. Basically, this method involves the strategic placement of windows in building design to capture the maximum solar energy potential.

peak oil The point at which the world has consumed exactly one-half of all the oil available and supplies begin diminishing. Also known as topping point.

per-capita energy consumption The average amount of energy consumed by each person in a country.

photovoltaics (PV) Solid-state semiconductor devices that convert sunlight directly into electricity, which can then be used for several types of applications. They require no moving parts or fluid, and they are either made of one of three types of silicon or come in the form of thin film made from other semiconductor materials. PV systems can be stand-alone, roof-mounted, or roof shingles.

power towers A type of central solar power system in which towers collect sunlight using fields of sun-tracking mirrors known as heliostats and concentrate the heat onto a centrally located receiver mounted on a tower.

precautionary principle A principle stating that, if the impacts of an action or inaction are extreme and irreversible, even if there is inconclusive scientific evidence about the risk for disaster, measures must be taken to prevent the impacts from occurring.

pressurized heavy water reactor (PHWR) Also known as the CANDU-A nuclear reactor; it was developed by Canada and only found there.

pressurized water reactor (PWR) The most commonly used nuclear reactor on the planet.

renewable energy Sources of energy considered to be inexhaustible. Such sources include the sun, wind, falling water, waves, tides, biomass, or heat generated beneath the surface of the Earth. Most renewable energy sources, except geothermal, are a type of solar energy.

renewable obligation certificates (ROCs) *See* green certificates.

renewable portfolio standard (RPS) A government requirement that sets mandatory goals for specific actors, such as utility companies, for renewable energy use and then lets the market determine the least expensive way to get there. The policy penalizes those who do not reach the goal.

small-scale hydro projects Small hydropower systems that capture energy through the use of small dams and river turbines. They often use the natural flow of the water to spin turbines or waterwheels so that they do not interfere with the flow of the river and do not usually include structures, like dams, that store water in reservoirs. Also known as micropower.

solar dish systems *See* parabolic dishes.

tapcon Also known as a tapered channel system. It is a wave energy technology that feeds into a reservoir located on a cliff above sea level. The narrowed channel causes the waves flowing inside of it to increase in height and spill over into the reservoir, where the water is used to run turbines in a manner similar to that of large-scale hydropower applications.

theory of technology development Theory that technology development occurs in stages. (1) In the conceptualization stage, the idea for the technology is conceived. (2) in the proof of concept stage, the idea is taken beyond the conceptualization stage, and a concrete model is created, often on a small scale, and then tested on a large scale until it is fully operational. (3) In the next stage, the design is tweaked until it is ready for market penetration. (4) The final stage is the market penetration stage, in which the technology is ready to meet the user's needs. Market penetration involves the commercial development of a technology that can be used on a mass scale so that it becomes available for widespread usage.

tipping point A concept borrowed from epidemiology in which small changes have no effect on a system until a threshold is reached. Once the threshold is reached, the effects are irreversible.

topping point *See* peak oil.

tradable renewable certificates (TRCs) *See* green certificates.

vertical axis wind turbine (VAWT) One of the two main types of wind turbines. It looks like an eggbeater.

wind power Wind is the result of differential heating of the Earth's surface. Differential heat is the difference between how much heat is absorbed by land and water surfaces. Wind energy is captured via a turbine, which turns as the wind blows. The turbine generates mechanical or electrical energy, which can then be used for a variety of applications.

Index

About the Authors

Zachary A. Smith is a Regents' Professor of Political Science at Northern Arizona University in Flagstaff. Dr. Smith has written more than 20 books and numerous articles on environmental and natural resource policy topics, and many consider him one of the premier experts in the field.

Katrina Darlene Taylor is a graduate student in the public policy PhD program, focusing on environmental policy, in the Department of Politics and International Affairs at Northern Arizona University in Flagstaff. Ms. Taylor is the coauthor of a chapter titled "Transborder Air Pollution" with Dr. Zachary A. Smith in *Handbook on Globalization and the Environment,* edited by Khi V. Thai, Dianne Rahm, and Jerrell D. Cogburn.